PARTICIPATORY RURAL APPRAISAL

Also available

Corruption at the Grassroots: The Shades and Shadows

About the Authors

N. Narayanasamy, Professor of Extension Education at the Gandhigram Rural Institute, has over 25 years of experience in teaching and research. Also, he is Programme Coordinator of the PRA activities at the Institute. His publications include: *Repayment Ethic, Extension in University Curriculum, Participatory Rural Appraisal, Analyzing Community Problems: Tentacles of PRA Methodology,* and *Corruption at the Grassroots: The Shades and Shadows.* He has also over 75 papers to his credit.

M.P. Boraian is Reader in Extension Education and Director of Employment Bureau at the Gandhigram Rural Institute. He obtained his Doctoral degree in Rural Development from Gandhigram Rural Institute and pre-doctoral degree in Rural Development from National Institute of Rural Development. He was a recipient of Fellowship from the United Nations' International Labour Organization. He has appraised, monitored and evaluated over 40 rural projects. He was associated, with Research Projects sponsored by UNICEF, ILO, SIDA, SAREC, Planning Commission and ICSSR. He has published more than 60 articles in professional journals at the national and international level.

PARTICIPATORY RURAL APPRAISAL

(The Experience of NGOs in South India)

Edited by

N. NARAYANASAMY

Professor and Head,
Department of Extension Education
Gandhigram Rural Institute
Deemed University
Gandhigram (T.N.)

M.P. Boraian

Reader
Department of Extension Education
Gandhigram Rural Institute
Deemed University
Gandhigram (T.N.)

CONCEPT PUBLISHING COMPANY, NEW DELHI-110059

Cataloging in Publication Data—DK
Courtesy: D.K. Agencies (P) Ltd. <docinfo@dkagencies.com>

Participatory rural appraisal : the experience of NGOs in South India/ edited by N. Narayanasamy, M.P. Boraian.
p. cm.
Includes bibliographical references (p.)
Includes index.
ISBN 8180691667

1. Participatory rural appraisal—India—Tamil Nadu. 2. Non-governmental organizations—India—Tamil Nadu. I. Narayanasamy, N. (Nammalvar), 1952- II. Boraian, M. P.

361.370 954 82 21

ISBN 81-8069-166-7

First Published 2005

Published and Printed by
Ashok Kumar Mittal
Concept Publishing Company
A/15-16, Commercial Block, Mohan Garden
New Delhi-110059 (India)

Phones: 25351460, 25351794
Fax: 091-11-25357103
E-mail: publishing@conceptpub.com

FOREWORD

Participation has come to occupy a predominant position in development research and work. It is being considered as panacea for all the ills of development. All the development experts and workers have come to an unanimous conclusion that development cannot be sustainable and longlasting unless people are made part and parcel of the development process. In order to make participation of the people in the development initiatives realistic and pragmatic, several approaches and methods have been discovered and rediscovered, tested and tried. Among them PRA stands out. PRA has emerged as an effective approach and method to empower the people especially the poor, the weak and the deprived. It is being extensively used in appraisal, analysis and action.

Realizing its efficiency and effectiveness, several development agencies and funding partners have put "PRA" as an essential component in different phases of the project. This has led to phenomenal spread of PRA. It is now used on a large scale and often at a very fast pace. It is practised in more than 100 countries. Such a quick spread, on the one hand, has created enormous opportunity for the application of PRA on a large scale; on the other, it has posed a serious threat to the very quality of PRA. Being largely qualitative in nature, the outcome of PRA is being questioned. It is against this background, a team of researchers from GRI, who have extensive experience in the practice of PRA, have undertaken a study on efficacy of participatory methods sponsored by Indian Council for Social Science Research, New Delhi.

Confined to NGOs practicing PRA in Tamil Nadu, the study provides a vivid picture on the state-of-art of PRA in Tamil Nadu. The study has proved that with proper training, adequate practice and constant applications, PRA could be used as an effective methodology in gathering information, project formulation, execution and evaluation. The study has further proved that PRA is less costly, less time-consuming and user-friendly. It helps in integrating the aspirations, needs and priorities of the people in project formulation.

It builds up the people's capacity, upgrades their analytical skill, release them from the shackles of dependency, mobilizes the marginalized people and empower the poor and weak. All these positive aspects of PRA have resulted in institutalization of the philosophy and principles of PRA in NGO's practicing PRA in letter and spirit.

The authors of this research-based book have ventured into an unexplored area of research. I congratulate Dr. N. Narayanasamy, Professor of Extension Education, and Dr. M.P. Boraian, Reader in Extension Education for their refreshingly analytical work. I strongly feel that the book would be of immense use and relevance to the teachers, researchers, NGOs, policy makers and donors who preach and practice participatory development.

Gandhigram Rural Institute
Gandhigram

Dr. G. PANKAJAM
Vice-Chancellor

PREFACE

Community participation commands topmost priority in the hierarchy of India's rural development goals. Rural Development is being redefined as participatory development. Seventy-third amendment to the Indian Constitution places people's participation in panchayats on top of the political agenda. In its development agenda, the Government, through its programmes and projects implemented both directly as well as through NGOs, assigns prime of place to the needs and aspirations of people. Donor agencies, bilateral bodies, multilateral institutions and United Nation's organisations too give people the leading role in the processes and programmes of development. People thus occupy a pivotal position in the development priorities of the Government, NGOs and Donor bodies.

The dawn of 21st century marks the birth of information era. No development process will be complete without gathering and sharing information from and with people. Adequate and accurate information is a vital requirement of every development pursuit. Correct, complete and timely data enables effective planning and efficient execution of development programmes for greater people's benefit. On the contrary, defective, deficient and delayed data will misdirect the course of development. In the process of gathering best information in a quick manner, the conventional social surveys are slowly losing their significance.

Participatory Rural Appraisal (PRA), the brain child of British Rural Development Expert Prof. Robert Chambers, is an innovative information gathering method used confidently by NGOs, convincingly by the Donor Agencies and carefully by the Government in select sectors. PRA has been used in almost every sphere, where surveys were used and PRA has been able to obtain every piece of information that surveys had elicited. Co-existence of social survey method and Participatory Rural Appraisal method led to testing their mutual strength and weaknesses, rather than establish their individual supremacy.

Participatory Rural Appraisal method could accomplish all that the social survey method accomplished and much more. On the contrary, social surveys, could not achieve what PRA achieved. Even where they concurred in terms of the output, the survey could not match PRA in respect of the process and outcome. PRA, thus, established itself to be the professional rival of social surveys, whereas social surveys could not prove to be so.

PRA proved to be not only different but also distinct from social surveys. Researcher goes to the people under field surveys, while the respondents come together and closer to the researchers. Survey is an interactive and information hunting process, whereas PRA is a collective and participatory process. While surveys elicit only information, PRA exercises generate interest, enthusiasm and confidence besides information. Survey is beset with the 'fag end fatigue' whereas PRA is confronted by full participation and active interest of people, throughout. Moreover, survey is merely an investigative process whereas PRA is a social process. Besides obtaining information, PRA also provides articulation, opportunity for participation, interaction, correction of ideas and above all a feeling of empowerment. Survey is merely a research tool whereas PRA is a tool for planning. PRA is, therefore, considered a shade better than a survey.

In the context of the above, it was sought to find out whether PRA is more efficacious and how the NGOs handled PRA, in what spheres, with what benefits, and botherations to them. ICSSR was kind enough to support this research project. The report provides answer to the above questions and highlights the practice of PRA among NGOs, comparative assessment of survey and PRA and the overall performance of PRA in the rural sector. The study is relatively a maiden attempt in the area of traditional and modern tools of research and planning, particularly in the non-Government sector engaged in the task of rural development. The researchers, therefore, welcome comments and observations on the methods and outcome of the study.

Gandhigram
April 2004

N. NARAYANASAMY
M.P. BORAIAN

ACKNOWLEDGEMENTS

We are grateful to the Indian Council of Social Science Research (ICSSR), New Delhi for sponsoring the research study on the "Efficacy of Participatory Rural Appraisal Methods: An Enquiry into the NGO Experiences."

We are thankful to our Vice-Chancellor Dr. G. Pankajam who is a constant source of inspiration in our research and extension endeavours. We record our deep sense of gratitude to Dr. T.R. Natesan and Dr. N. Markandan our former Vice-Chancellors who encouraged and supported us in all our PRA related work. We are particularly thankful to Sri D.K. Oza, former Vice-Chancellor, who introduced us to the world of PRA.

We record our thanks to the authorities of our Institute for providing necessary financial assistance for publishing this book. We also thank our Registrar Dr. P. Kanniappan and his predecessors who extended necessary cooperation in this pursuit.

We express our sincere gratitude to all the heads and staff of Non-Government Organizations who responded to our study favourably, by providing the required information for the study, in the process of practicing PRA in their service areas. Prominent among these NGOs are: OUTREACH, SPEECH, ASSEFA, WOMEN'S COLLECTIVE, and RENAISSANCE. In these organizations, we owe our indebtedness to Messers. James Mascarenhas, Radhakrishnan, and Arulsamy, OUTREACH, Messers. Erskin, Nirmal, Venkataraman, Karuppiah, SPEECH, Mr. A. Arockiasamy, ASSEFA, Messers. Sheelu, Lidwin, Selvi, Bama, Usha, Sunitha, Viji, and Arul, WOMEN's COLLECTIVE, Mr. M.S.T. Chandrasekar, RENAISSANCE who shared their rich expertise in using PRA.

We thank Sri B. Baskar, Research Assistant and Ms. V. Jerinamary, Secretarial Assistant who provided field and office support in collecting data, processing and computer feeding the report.

We are indebted to Dr. S. Manivel, Lecturer in Cooperation, Gandhigram Rural Institute for helping us in various forms for the conduct of the study.

Gandhigram
April 2004

N. NARAYANASAMY
M.P. BORAIAN

CONTENTS

Foreword by ***Dr. G. Pankajam*** v

Preface vii

Acknowledgements ix

List of Tables xii

List of Exhibits xiv

Acronyms and Abbreviations Used xv

1. Participatory Rural Appraisal: Genesis, Principles and Methods 1
2. Design of the Study 21
3. State of Art of PRA among NGOs in Tamil Nadu 46
4. Use of Participatory Methods in Project Cycle: The Experiences of NGOs 88
5. Efficacy of Participatory Methods 145
6. Summary of Findings and Policy Implications 178

Bibliography 196

Index 206

LIST OF TABLES

3.1 : Years of service of the NGOs 49
3.2 : Objectives of the NGOs 51
3.3 : Sector-wise activities undertaken by NGOs 53
3.4 : Jurisdiction of Non-Government Organisations 55
3.5 : Beneficiaries/clients of the NGOs 56
3.6 : Methods followed at different stages of the project cycle 57
3.7 : Problems faced in formulation of projects 59
3.8 : Problems faced in implementing the project 60
3.9 : Problems faced in monitoring & evaluation of the project 61
3.10 : Measures taken to overcome the problems in project formulation, implementation, monitoring and evaluation 61
3.11 : Years of experience of NGOs in PRA 63
3.12 : Number of training programmes undergone by staff members 64
3.13 : Duration of training programmes undergone by staff members 64
3.14 : Number of PRA methods used by NGOs 66
3.15 : Use of participatory methods by NGOs 67
3.16 : Application of PRA by NGOs 69
3.17 : Benefits derived from using PRA for the outsiders 71
3.18 : Benefits derived from using PRA for the people 73
3.19 : People based problems 75
3.20 : Team based problems 77
3.21 : Approach based problems 79
3.22 : Tool based problems 80
5.1 : Efficacy of PRA as reported by NGOs 153
5.2 : Time taken to gather data: a comparison between survey and PRA methods 158
5.3 : Cost of data gathering: a comparison between survey and PRA methods 162

5.4 : Comparison of data collected through social mapping and survey method 164
5.5 : Participation in social mapping at Chellakuttiur village 169
5.6 : Participation in social mapping at Pannaikulam village 171
5.7 : Participation in Social mapping at Karuthakkapatti village 172
5.8 : Extent of active participation in social mapping village 173

LIST OF EXHIBITS

1 : PRA in institution building 92
1A : Venn diagram 94
2 : Trends 95
3 : Usthanapalli: Seasonal calendar 96
4 : Mandinangam: Resource location and mobility map 98
5 : SWOT analysis on proposed agro-service centre 100
6 : Self evaluation of Lalitha - after joining Sangam 103
7 : Major expenditure in a year 105
8 : Reasons for the immediate sale of groundnut after harvest 107
8A : Groundnut cropping pattern 108
9 : Self-monitoring of Kamakshi Mahalir Sangam 109
10 : Ranking of bad qualities in a family 111
11 : Ranking of bad qualities in a family-chappathi diagram 112
12 : Impact diagram - Ladder scale 114
13 : Innovation-assessing the role of VFC members and other users 114
14 : 100 seeds scoring 115
15 : Anxiety map through family counseling technique 116
16 : Development strategy of SPEECH 120
17 : Village organisations - linkages 127
18 : Logical framework analysis 130
19 : Impact assessment 135
20 : Evaluation of project area 137
21 : Evaluation of project area - women's status 138

1

PARTICIPATORY RURAL APPRAISAL
Genesis, Principles and Methods

Introduction

The past five decades witnessed the evolution of many development strategies and theories. Of them, two main trends have emerged in practice.[1] The first is where 'development' starts with the outsiders providing some goods or services, which a community may or may not need. This approach is fundamentally about the delivering of resources which come in various guises like finance, equipment, technical know-how, skills, even a particular approach to life, etc. This, also known as top-down approach, is still prevalent. The second one is people-centered or bottom-up approach to development. This approach concentrates on the needs of the people, what they want and need. They define the goals of development and participate in development 'projects' from the beginning.[2]

Top-down approach: Assumptions

The top-down approach is based on certain assumptions and practices.[3] They are: (i) development can be created and engineered; (ii) it is something which is brought, to and for some, by others who presumably are more developed; (iii) it is done on behalf of third parties and not by the communities and clients who are the subjects of the intervention; (iv) it is linear and predictable. In other words there is direct link between cause and effect; (v) it presumes that understanding will generate change; (vi) it places far more emphasis on technical experts and 'advisors' and on trainers, than it does on change facilitators; (vii) it assumes a preferred culture or value system. The presumption is that there is something wrong and we intervene to change it; (viii) it sees participation as a means, not as an end it itself; (ix) it assumes that a successful development project is replicable and sustainable; and (x) the development project is evaluated with reference to output and not to their outcomes.

Development Process

The assumptions of top-down approach have been questioned on the ground that development is an innate and natural process found in all living things. Development cannot be brought or delivered. The development process already exists. The development practitioners just intervene to facilitate the process of development. This necessarily presupposes that the development practitioner has to understand the development process into which he or she is intervening. It is important to know where the individual, the organization or the community is located on its own path of development. It is equally important to understand where it has come from, how it has changed along the way and what the next development challenges are likely to be. All these in turn require openness; an ability to observe accurately and a fine understanding of development process.[4] The top-down approach has failed to recognize the fundamentals of development process. As a result, the approach could not make any desired impact on the life of the people.

The bottom-up approach has totally pinned its faith on the knowledge of the people and used participatory strategies at different stages of the project though the degree and level of participation vary from approach to approach. The bottom-up approach identifies three discernible phases of ideal unimpeded development. The first phase characterized by dependence is a period of great learning and skill acquisition in which both the outsiders and the people play a major role in providing the environment and resources required for growth. The second phase of independence entails a fundamental change in relationship and a period of testing and personalizing capacities and competencies, using them to act and impact on the environment in ways that help establish the people as unique and self-reliant. The third phase involves another fundamental change in relationship towards increasing inter-dependence—the people now understand that the full realization of potential is achieved only through effective collaboration with others.[5] All these three phases of development are equally important and people must find a prime place in all the three phases.

Participatory Approaches

Several methods, strategies and approaches have emerged during 70s, 80s and 90s placing people as the prime actors in the process of development. All these methods emphasized the participation of people at different stages of the development process. Many of these methods have richly contributed for the evolution and growth of PRA. Some of them stand out. They are :

- Participatory Action Research
- Agro-Eco-System Analysis
- Applied Anthropology
- Informal Survey
- Rapid Rural Appraisal

The idea behind each one of the above approaches, the purpose and methods followed under every approach and contribution of each approach for the genesis of PRA is briefly presented.

Participatory Action Research (PAR)[6]

Participatory Action Research (PAR), is a development strategy in which the community participates fully and through its own critical analysis, gets to know the facts of its own situation, condition, strengths, weaknesses, resources, problems, etc. The process of PAR involves an effort to generate and consolidate the community's knowledge, skills and practices that are present among the poor and the marginalized which in turn could be used as a powerful tool for their empowerment and liberation.

PAR is used to collate old facts and empirical data from the people—mainly the marginalized and also to generate theories relevant to specific social situations. The people are the prime movers in the participatory action research and they do every thing. Their participation takes place from the very day the research begins. It is a collective effort by the participating community to pool their efforts and other resources to achieve the objectives that they have set for themselves.

In short, Participatory Action Research (PAR) is both a methodology and a process, wherein people are involved in actions/activities that have been planned, designed, implemented and controlled by the community for its own betterment. The major goal of PAR is to rediscover, release, restore, renew, empower and transform the people's situation/condition so that they have control over their own situation.

PAR aims at

- empowering the weak and the poor so that they are enabled to lead a fairly good life with what they know and what they have;
- enabling the members of the community especially the oppressed and the marginalized to come together and share a common platform, discuss and analyze common issues and deciding on an workable action plan;
- making the people to realize their self worth, their potentials and usefulness of their role in society; and
- building-up self-reliance in the village community and thereby relieving them from dependency culture.

PAR believes

- in the fundamental intelligence of people irrespective of whether they have had formal education or not, as they may be having immense experiential knowledge.
- that everyone including the poorest has not only the right to make history but also the right to create knowledge and formulate theories.

— that the marginalised/exploited poorest with their experience and knowledge could become resource-rich people.
— that everyone is capable of learning.
— that the people learn best by doing.
— that knowledge is power.
— that success helps in building confidence and makes people realize their self-worth.
— in the strength of collective learning and in collective action.
— that when people are organized they are able to recognize their own capacities and are better equipped to counteract abuses and disrespect.

The contributions of PAR to PRA are as follows

— the professionals should reflect critically on their concept, values, behaviour and methods.
— learning through engagement and committed action.
— professionals have roles as conveners, catalysts and facilitators.
— the weak and the marginalized can and should be empowered.
— instilling in the minds of the rural especially the poverty-stricken, that they can and should involve in investigating and analyzing their own problems as well as plan for solutions.

Agro-Eco-System Analysis[7]

The agro-eco-system can be defined as an ecological system partly modified by man to produce food, fiber and other agricultural produce, which is useful for the mankind.

The system permits both for experimental and applied research for the purpose of its continued development and sustainability.

Agro-eco system has four different but interconnected system properties. They include :

Productivity: It is defined as "the net output of valued product per unit of resource input. Product is measured as field or net income on food value etc. The resources may be land, labour, capital, energy, fertilizer, pesticides etc. Productivity, therefore, may be with reference to land, labour, capital, technology etc.

Stability: It is the constancy of productivity in response to small disturbances or changes caused by the normal fluctuations of the surrounding physical, biological, social and economic environment.

Sustainability : It is the ability of the farm ecological system to maintain productivity and stability inspite of major environment disturbances such as stress and shock which can neither be predicted well in advance nor measured accurately.

Equitability: It is the evenness of distribution of the production of the agro-eco-system among the human beneficiaries.

The Agro-eco-system develops through a process of intensification, which normally involves :

- transformation of land use
- existing land use

The intensification process for development of agro-eco-system may have positive as well as adverse consequences. Therefore, it is important to trade off between different elements in the agro-eco system.

Agro-eco-system analysis developed in Thailand in the late 70s at the University of Chiang Mai by Gardon Conway and his colleagues, spread first through Southeast Asia and then to the other parts of the world. Conway worked further with others especially with Aga Khan Rural Support Programme (Pakistan) to develop the agro-eco system analysis more suitable for identification and assessment of practical hypothesis for action in the field of agriculture.

The agro-eco system analysis makes use of several tools. Some of the major contributions of Agro-eco-system analysis to PRA are :

- Transects (systematic walks and observation).
- Informal mapping (sketch maps drawn on sites).

- Diagramming (seasonal calendars, flow and causal diagrams, bar charts, venn diagram).
- Innovation assessment (scoring and ranking different action).

Applied Anthropology[8]

Traditionally, anthropology has dealt more with understanding of the facts of human existence than dealing with the solutions to human problems. However, 1970's and 1980's witnessed the establishment of anthropological approaches in a professionalized manner, sorting out the vicissitudes of social change in tribal and rural communities.

The present repertoire of PRA owes a great deal to the insights gained from the field of applied anthropology. To emphasize some of them :

- Recognizing that learning from the field involves the flexibility contained in arts more than the rigidity seen in a scientific endeavour;
- The meaning and value of learning from relaxed and unhurried interactions and observations with people in their natural settings;
- Importance of establishing rapport with the people with the important role of the underlying attitudes and behaviour;
- The emphatic emic-etic distinction; and
- Recognition of richness and validity of indigenous technical knowledge.

Informal Survey or Sondeo[9]

Agricultural research teams, after seeing the "expensive, bulky and hard-to-process questionnaire survey, have gone in for informal survey. The informal survey is a data gathering method which can produce at minimum cost a rich description of life in a farming community, an understanding of local ecology, cropping systems and how farmers, merchants, extension workers, and others perceive their conditions and make decisions.

Informal survey can provide basic data and information on the feasibility of beginning a project in a region. This is so especially when dealing with farming system about which little is known. It can also be used to quickly obtain basic information specifically for the design and execution of formal surveys or more in-depth investigations. The important purpose is to help focus a subsequent formal survey that will utilize random sampling and quantify critical aspects of rural life or the production system. The need is not simply to get a feel of the area, but to discover important, albeit tentative, organizing concepts upon which to base future research.

The informal survey is methodologically simple but physically tough. The successful survey may require sloshing through muddy fields and scrambling along rocky paths and dangerous slopes. The surveyor must be country-oriented, grabbing out information in the fields, market places or wherever farmers' daily routines carry them.

The successful informal surveys also require mental and methodological flexibility. It does not proceed like the formal questionnaire survey where pre-determined hypothesis is tested. Instead, important questions and direction of study emerge as information is collected. This is not to say that formal survey lacks logic but that one must be able to accommodate new information and adjust research plans accordingly. As the survey advances, the team will pass initial vagueness to a mid-way focussing and finally arrive at a stage where threads can be pulled together and specific ideas tested.

The major contributions of informal survey to PRA are :

- Observation—the researcher needs to keep the eyes open for patterns in crop production, land use and farm behaviour.
- Informal conversation—the researcher needs to talk with the people and listen to their concerns and views.
- Inter-disciplinary team
- Recording—the researcher has to discretely write everything down. Complete field notes are crucial.

Rapid Rural Appraisal[10]

The Participatory Action Research System and Agro-Eco System analysis were developed and experimented in the 70s for the purpose of gathering data and information for quickly appraising the situation and conditions of the farmers and the agriculture especially in the third world countries. The methods adopted by the experts for agro-eco system analysis did not, however, find their place in fields other than agriculture. But the development efforts and strategies are not confined to agriculture alone. It has a wider perspective and covers a whole range of activities, which affect the life of people in a locality, or country. Therefore, the development professionals felt that it is essential to have better understanding of the overall situation of the local area to plan, execute and to evaluate the development and welfare programmes. Therefore, they started using quick appraisal methods to appraise the conditions for planning their programmes.

Rapid Rural Appraisal is a way of organizing people for collecting and analyzing information within a short time span. It can be defined as any systematic process of investigation to acquire new information in order to draw and validate inferences, hypotheses, observation and conclusions in a limited period of time.[11] It has flexibility to adjust to situations because it does not imply or recommend a standard set of methods to be applied in each case. The methods vary from situation to situation and are determined by local conditions, local problems and objectives at hand.

RRA, as Robert Chambers[12] has put, had three main origins. The first was dissatisfaction with the biases, especially the anti-poverty biases of rural development tourism (the approach of brief visit to rural areas by the urban-based professional to understand the socio-economic conditions of the people in the rural areas). Second, the disillusion with the normal processes of questionnaire surveys and their results. Third, the continuous search for most cost effective methods of learning the rural conditions.

In the evolution process of the RRA approach, certain principles emerged and these principles describe the RRA as a method of appraisal of a situation largely. The RRA approach can be described as a semi-structured way of learning, relatively quickly and in a multidisciplinary team, from local people about the key problems and opportunities of an area and of deciding on an agreed set of possibilities for alleviating the problems and/or making good use of the opportunities.

The RRA approach thus aims to :

- allow the outsiders to learn from the local people,
- go beyond and get answers/responses, which the formal surveys can not get,
- challenge anti-poverty biases, which are common in other approaches especially in the Development Tourism,
- be quick, flexible and adaptable to the rural situations,
- get away from roads and reach interior villages and remote places,
- avoid over collection of data and collecting only the required data, and
- accept a degree of inaccuracy.

Advent of Participatory Rural Appraisal (PRA)

The approaches discussed have some features in common. They may be listed: (i) they are fairly rapid than conventional methods; (ii) all are field-based; (iii) the emphasis is on learning directly from the people; (iv) they avoid blue-print approach and adopt semi-structured multi-disciplinary approach; and (v) focus is more on the process rather than on outcome.

Yet, all these approaches remain extractive in nature, as people are not consulted beyond the phase of data collection. In other words, participation of the primary stakeholders (people) of the development project ends with collection of data and information. They are not involved in the analysis of their situation. This was considered as a serious lacuna in the participatory field-oriented approach. Hence, the emphasis shifted from rapid to 'participatory' and empowering, which fitted well with participatory approach to rural development and become the starting point of PRA.

PRA: Meaning

It is a family approach and a method of enabling local people to analyze their living conditions, to share the outcomes, and to plan their activities. It involves "handing over the stick from outsider to insider" in methods and action. The outsiders' role is catalytic, as a facilitator and convenor of processes within a community that is prepared to alter its own situation.

Basic Components of PRA

The three basic components of PRA are facilitation, attitude, and behaviour of outsiders and culture of sharing. In PRA, outsiders act as facilitators and enable local people to do all or most of the investigation, mapping, modelling, diagramming, analysis, presentation, and planning. Similarly, the facilitators are expected to follow the local norms to the extent possible so that local people can feel they are part of their own community. This requires the outsiders or facilitators to pay attention to their every activity such as where to sit, how to listen to local people (rather than lecturing) appropriate visiting time, respecting local people's knowledge, skills and expertise, taking an attitude of learning and embracing error. Finally, continuous sharing of information with local people, with other PRA team members, and with other institutions, NGOs and government, is another basic component of PRA that has to be considered by the PRA team members all the time.

Principles of PRA

PRA is based on certain principles. They are revolutionary in nature though they are not declared from any pulpit of revolution. They constantly question the conventional modes of understanding rural problems, and have evolved certain principles, which are easy to follow and adopt. The are stated below.

Reversal of Learning : In PRA, there is a total change or shift in the learning process. PRA encourages the novice outsiders to accept their ignorance about rural realities and be willing to learn from the rural people. It believes in flexibility and openness and not in a blueprint. It underscores its faith in learning through and with group. It pins its faith in empowering the people and not in extracting information from them. Its approach is holistic and not reductionistic. It has made the process of learning a fun by using a mix of different methods and techniques that enable the people to participate at every stage of the development process.

Learning Rapidly and Progressively: The learning process is flexible and adaptable. The methods are improvised and employed depending on the local situations. There is ample opportunity for active interaction with the people and also for exploring and identifying undiscovered areas in development practice.

Off-setting Biases: The different types of biases found in conventional ways of learning are off-set by listening, probing, being unimposing, deliberately seeking out the neglected and marginalised and learning their concerns and priorities.

Optimizing Trade-off: This includes the principle of optimal ignorance and the principle of appropriate imprecision. Optimal ignorance means avoiding over collection of data. Whereas appropriate imprecision means not measuring more than needed.

Triangulation: It is carefully cross-checking the information gathered by using a mix of techniques, by employing multi-disciplinary team and by turning to different sources of information.

Seeking Diversity: PRA believes in variability and not in averages. The purpose is to maximize the diversity and richness of information. It deliberately looks for, notices and investigates contradictions, anomalies and differences.

They do it: PRA has a strong faith in peoples' knowledge and capabilities. People are facilitated to do everything related to data gathering, analysis, interpretation, drawing inferences, prioritizing the problems, suggesting solutions or alternative courses of action. The outsiders are mere facilitators, learners and catalysts.

Self-critical Awareness: This is a very cardinal principle of PRA. The outsiders, as facilitators constantly examine their attitude and behaviour and try to do better. They embrace the errors committed by them in the appraisal process. They also welcome error as an opportunity to learn better and do better. The golden rule is "In all circumstances use one's own best judgement". This would mean accepting personal responsibility rather than looking for manuals or a rigid set of rules.

On the Spot Analysis: The data and information gathered are analyzed on the spot by the people facilitated by outsiders. Analysis at the office is always avoided. On the spot analysis of data and their interpretations by the people often astonish the researchers with new revelations.

Sharing: The information gathered are shared not only with the people but also with others who are interested. Sharing is the fundamental rule, which is practised by everybody involved in the appraisal. Sharing would also mean sharing of even food sometimes.

Action: It is good to make it a point never to leave the village high and dry after the situation appraisal or planning process is over. The closeness of working relationship and the affinity PRA develops between the outsiders and insiders, definitely make the rural people to expect some action programme to better their living conditions. Moreover, rural people have cooperated and participated all through the PRA process. So, it is always good to initiate some kind of joint action.

Methods of PRA

PRA uses a wide range of methods. They are not new nor inventions. Many of them are borrowed from different disciplines. They are improvised and used depending on the local situation and condition. The practitioners of methods should be flexible and innovative in using the methods. The golden rule is "use your best judgement in all circumstances". The facilitator must be thorough with the method he uses. He must be in a position to explain the method to the people in a simple, lucid style so that the people are able to understand and use the methods in the field without difficulties. In other words, methods must be mutually understood. There are more than 50 PRA methods. We present a few methods, which are often used in the field.

Social, Resource and Land Use Map : Mapping as an exercise in PRA involves a group process in which a sketch of the community is compiled in cooperation with a group of local men and women to identify physical and socio-economic details along with the infrastructure available in the community. Depending upon the purpose of the exercise, different names can be given to such sketches such as a social map, resource map, land-use map, etc. The aim of mapping is to allow local people to express their perceptions of location, usage patterns, and changes of local resources or facilities. Maps are the excellent source of information sharing. Its visual character attracts greater participation of people/villagers in the PRA exercise.

Transect: Transect is a systematic walk with a purpose along with a few key informants through an area. The aim is to have a very clear understanding of the locality by identifying different zones, observing, asking, listening, discussing local technologies, problems, seeking solution, opportunities, and mapping and diagramming the resources and findings. This technique has the advantage of leading to field-based observations, which can be discussed with local people in the subsequent sittings.

Seasonal Calendar: This tool is applied to collect seasonal information such as intensity of rainfall or soil moisture, land use or cropping patterns, migration patterns, food availability, income and expenditure patterns by month. Local people use different lengths of sticks or straw along with counters (e.g., seeds) to chart on the ground the relative quantities of some variables. Seasonal calendars also help record villagers' views of problems and opportunities.

Time Line: Time line is a record of events and activities, which occurred in the community in the rememberable past. The events and activities are, by and large, are recorded in a chronological order. This exercise involves discussion with a group of local people (usually 4-8) regarding what they consider to be the most important past events. This is a good ice-breaker for building rapport with local people as it shows an interest in their lives. The main purpose of the time line is to identify events in time to which local people can refer when discussing historical issues.

Trend: Trend analysis attempts to study people's accounts of the past, of how things close to them have changed at different points of time. Information sought through the technique of 'trend' gave insights on various facts such as ecological histories, changes in land use and cropping patterns, changes in customs and practices, changes and trends in population, migration, fuels used, education, health, credit.... and the causes of changes and trends could be ascertained using trend analysis method.

Trend Line/Diagram : Trend lines are developed according to village perspectives to show patterns of changes along with the causes of such changes in resources and issues such as rainfall, crop production, soil loss, deforestation, livestock holding, and other topics of concern to the community. A group (15-20) of local people representing different age groups, knowledgeable about the topics to be explored could be gathered for this exercise.

Activity Profiles and Daily routines : Activity profiles and daily routines are exercises, which help exploring the daily patterns of activity through profiles and routines. Chart for each hour of the day with typical activities, amount of effort involved, time taken, location of work etc., are elicited and used for understanding the routines of the different classes, age groups, sexes of people in a community. It also allows comparison between and among different people e.g., men, women, old, young; compare profiles and routines for different seasons.

Local Histories : Local histories are the detailed accounts of the past, of how things have changed, particularly focusing on relationships and trends. These include technology histories and review, crop histories, livestock breed histories, labour availability, trees and forest histories, education change, and population change. Folklore, songs and poems are valuable resources for exploring history.

Local Researchers and Village Analysts : People concerned with the development of the community are well versed with the local issues and problems. They also attempt to analyze the issues and problems in detail and in their effort find solution to such issues. The knowledge of such people could be a great source of information for an outsider to learn and to the villagers to share. Even during the PRA exercises more and more local people are taking an active role in analysis and presentation. Village extensionists conduct complete participatory analyses with no outside help; villagers conduct transects, interview other villagers, draw maps, observe and produce plans. Other potential researchers include school teachers, students, poor etc.

Key Probes : Key probes are questions that can lead directly to key issues, such as "what do you talk about when you are together?"; "what new practices have you or others in this village experimented in the recent years?"

Sequencing and Chains of Interviews : There are many types of interviews that may be combined in sequences and chains.

Key Informant Interviews : Arranging for a series of interviews with key informants of the different stage of a process (e.g., men on ploughing, women on transplanting and weeding, shopkeeper for credit and inputs etc.).

Group interviews : These may be group convened to discuss a particular topic (focussed or specialist groups); groups comprising a mix of people whose different perceptions illuminate an issue (structure groups); casual groups, say, at a tea shop; community and neighbourhood groups. Group interviews provide the focus for discussion. Group interviews are often powerful and efficient.

Inventory of Local Management Systems and Resources: Local people know their area best, and have evolved their own systems of management. Use local classifications wherever possible—these are often more diverse than those used by outsiders. Every village has its local taxonomy and classification.

Matrices: Matrices are useful for ordering and structuring information gathering and planning. They include :

- attributes matrices for technologies e.g., dams and catchment structures along the top and attributes down the side;
- problems-opportunity matrices, with column headings including, for example, local name for zone, local name for soil, soil type, cropping pattern, problems, resources available in village and lastly, possible solutions;
- manual discriminant technique matrices for contract comparisons. Ask group A why group B is different or does something different. For example, set up three groups of high, medium and low maize yielding farmers, and ask why they get the yields they get, and why the others get their's. Record the key findings in a matrix.

Matrix Ranking : Matrix ranking helps to elicit information about local people's preferences with regard to tree species, types of livestock, varieties of crops, etc., and the criteria on which those preferences are based. While the criteria are listed to the left, the preferences that are to be compared with one another are listed on the top.

Matrix ranking is best done by two people—one to ask the questions and conduct the interview, the other to keep notes and collate and list the criteria. The second person can also observe what goes on in the group, note potential key informants for follow-up and list points for further enquiry.

Direct Matrix Ranking (DMR): Direct Matrix Ranking describes the method because it moves quickly from early discussion and questioning to recording respondents' views directly onto a table or matrix. It is simple, quick, and informative, and everyone seems to learn something from it.

Wealth Ranking: Wealth ranking is a technique that can be used for gaining rapid insight from a local perspective into factors affecting differentiation. It combines in-depth discussion of wealth with a ranking exercise that allows the participating group of local people to assess the relative wealth of households in a preselected list.

This method helps categorize households according to wealth or well being in the community. Key informants first develop the parameters they think are important to consider while ranking households and then keep placing households into appropriate ranks based on those parameters.

The technique has a number of potential applications :

- The stratification of a sample according to wealth criteria for further focussed appraisals on particular sub-sectors of the population.
- The generation of questions for further research into rural differentiation.
- An examination of survey data from a farmer's perspective.

Venn or Circle Diagramm : Venn or circle diagrams are used as a tool to discuss the relative importance or position of different factors, commonly institutional or social structures, and their closeness to the people in a community. Key informants are asked to rank community institutions in order of importance and to construct diagrams that indicate the relationships between and among village units. Circle of different sizes and colours represents organizations, institutions, or prominent people. Their relationships to each other and relative importance in the community can be mapped out by placing these circles on the ground in relation to each other.

Semi-structured Interviewing : This technique, also known as informal discussion, is considered the core of good PRA. It is a kind of open discussion with open-ended questions, which can take place anywhere in the community, either with individuals and/or groups of key informants. It can be done with mental or written checklists. These conversations can take place on the path while observing community activities, over the garden fence, and in fields or homes.

Night Halt : All interactions between outsiders and villagers are facilitated by staying in the village—the night halt. Many organizations have made this an essential part of participatory analysis and planning. It provokes change in outsiders' attitudes—they sleep and eat as villagers do; it allows for early morning and evening discussions when people are less busy; and it is an explicit indication of commitment by the outsiders and brings the outsiders very close to the people in the village, which strengthens mutual understanding.

Self-correcting Notes and Diaries : This is not of course a PRA tool which enables sharing information, but it aids in a better learning process. It helps to a greater extent to improve the outcome of the PRA exercises. The PRA practitioner by keeping a private diary or series of notes to focus on where he, as an outsider, would desire things to go better next time? Where were the problems? What could be done to avoid them? Who might be able to provide some solutions?

Shared Presentations : The findings are presented back to villagers and outsiders. It would be better if villagers do it themselves. It offers opportunity for cross-checking and feedback. Outsiders stand up, present, and invite comment and criticism—a reversal of roles. It is essential that outsiders concentrate upon both presentation and analysis of who contributes. The findings are presented back to villagers and outsiders. Who talks? Who interrupts whom? Whose ideas dominate? Who is different from everyone else? Who lectures? Outsiders must concentrate in listening and facilitating. Not lecturing.

Conclusion

Participatory Rural Appraisal, as an approach and as a method, has primarily emerged to strengthen, reinforce and sustain the participation of the people in development intervention. The ultimate aim of PRA is self-mobilization and empowerment. The aim, the advocates of PRA felt, could be realized as the concept of PRA is simple; as the principles are mostly ground rules and pragmatic; and as the methods could be practised even by naive villagers. But then, the outsiders or facilitators need to play a very significant role for grounding the PRA at the grassroots and also in enabling the participants to practice the participatory methods with all its essential features. Thus, both the outsiders and insiders have equally a major responsibility in realizing the aim of the PRA. Their role in tandem determines the efficacy of participatory methods. The focus of the thesis is on efficacy of participatory methods. This chapter provides a prelude for understanding the inferences drawn on subsequent chapters.

NOTES AND REFERENCES

1. Marian Meyar and Naresh Singh, "Two Approaches to Evaluating the Outcomes of Development Project", *Development in Practice*, Vol. 7, No.1, Feb. 1997, p. 59.
2. Goulet, "Authentic Development: Is it sustainable?" in *A Sustainable World: Defining and Measuring Sustainable Development* (Trzyna and Osborn, Eds.) Sacranento : International Centre for the Environment and Public Policy, California Institute of Public Affairs, 1995, p. 212.
3. Allan Kaplan, *The Development of Capacity*, NGLS Development Dossier, UN Non-Governmental Liaison Service, Geneva, 1999, pp. 5-6.

4 *Ibid*, p. 11.

5 *Ibid*, p. 12.

6. Hari M. John, "In True Partnership with people : Participatory Action Research and the Poor", *SEARCH News,* Vol. VII, No. 1, Jan.-March 1992, pp. 8-13.
7. Gardon R. Conway, Jennifer A. McCracken and Jules N. Pretty, *Training Notes for Agro-eco-system Analysis and Rapid Rural Appraisal,* IIED, London, 1987, pp. 5-18.
8. Narayanasamy, N., Dwaraki, B.R., Boraian, M.P., and Ramesh, R., *Analysing Community Problems: Tentacles of PRA Methodology,* Himalaya Publishing House, Mumbai, 1998, p. 12.
9. Robert E.Rhoads, "Basic Field Techniques for Rapid Rural Appraisal", Proceedings of the International Conference on Rapid Rural Appraisal, 1987, pp. 114-128.
10. Robert Chambers, *Whose Reality Counts? Putting the First Last,* Intermediate Technology Development Publications, London, 1997, pp. 110-113.
11. Neela Mukherjee, *Participatory Rural Appraisal: Methodology* and *Applications,* Concept Publishing Co., New Delhi, 1994, p. 27.
12. Robert Chambers, *op.cit.*, p. 111.

2

DESIGN OF THE STUDY

Introduction

Participatory Rural Appraisal, popularly known as PRA, is a novel method of ascertaining the problems facing the community, groups and individuals in a village, assessing the available natural and physical resources and identifying commonly acceptable solutions for formulating time-bound action plans for possible implementation, primarily with their own resources and wherever feasible, with those mobilized from outside the community.[1]

The origin of PRA, as a methodology, can be traced back to the multitude of methods and approaches which gradually emerged since the seventies, mainly due to poor results obtained from application of the conventional methodology to rural development. Important among such methods that emerged as an alternative to conventional methods[2] are :

- Rapid Rural Appraisal or Rapid Appraisal.
- Exploratory Survey.
- Informal Agricultural Survey.
- Agro-ecological zones.
- Agro-eco system analysis.
- Rapid agro-eco system zoning.

All these methods are field-oriented and have some features in common. They are quicker than conventional methods, applied in field situations, learning directly from people, semi-structured multi-disciplinary approach with flexibility and innovation, emphasis on understanding the process through involving people and gathering insights on problems and their probable solutions rather than final answers and recommendations.[3] However, RRA and similar methods remain extractive, whereby the researchers observe, write down and analyse. The field-oriented methods leave much scope for inducing greater element of participation.

The requirements of the shifting paradigm of development was not only a relatively quicker method of data collection but also one which would help local people to perform their own analysis. Hence, certain elements were required to be incorporated in PRA and other field methods to make them participatory and more broad-based in terms of their application. This necessitated the shifting of emphasis from rapid or quick to participatory and empowering. This has resulted in the birth of PRA.[4]

Statement of the Problem

PRA has been spreading very quickly. Practised earlier by NGOs, it has now been practised almost by all who are working in the field of Rural Development. PRA in many development projects has now become a 'flag of necessity'. Primarily used for appraisal of rural potentials and problems earlier, leading to follow-up action through formulation of action plans, PRA has now been used by a variety of audience in varied settings for variegated purposes. The concept of PRA has, therefore, been growing and expanding, its application has been widening and broadening, its popularity has been rising and resounding and its need has become stronger and severer. Thus, in a short period of time, PRA as a methodology, has gained strength through its emphasis on participation in line with the paradigm shift which is slowly taking place and has helped to fill the vacuum existing in the sphere of rural development.

Rapid spread of PRA has also resulted in criticisms against PRA. It has raised certain fundamental questions against the very outcome of PRA methods wherever used. With their vast and varied experience in practising and propagating PRA over a period of more than a decade in different socio-cultural settings, practitioners of PRA have sensed that the second generation of PRA users have started using PRA just for the sake of it as a fashion of the day. Practice of PRA for many of them is "we also do PRA". Such an attitude could result in neglecting the basic philosophy and principles of PRA, which in turn, could lead to lesser degree of participation and poor quality of information.[5]

Further, stakeholders' participation has become a requirement in most donor supported projects, and is stressed more and more in government programmes too. This has led to some good results. Much more common though, has been its abuse and bad practice. This has occurred on a huge scale, resulting in degeneration. The causes for the degeneration could be more specifically listed. They are: the rush to prepare projects and programmes; drives to disburse funds; time bound targets for producing reports, neglecting process; neglect and underestimation of the knowledge and capabilities of local people; neglect of local capacity building and institutional development; and failure to recognise personal and professional attitude and behaviour change required for PRA.

There is also real danger of PRA becoming a routine in which a set of methods is pursued almost like a drill, missing the central point, resulting in little attitudinal change and very superficial layers of information. There is some substance that participatory methods constitute an inquiry that is undisciplined and sloppy. They are said to involve only subjective observations and so respond fast to selected members of communities.[6] These criticisms deserve attention and the PRA practitioners should adhere to self critical evaluation. Otherwise, participatory methods of dealing with rural people may get into a rut without any scope for conscious exploration or concrete development.

The questions, under such a situation, need to be addressed are: What is the present status of PRA? What has been the experience of NGOs? How do they view PRA as a methodology of gathering information, for formulating project, planning their activities, executing and evaluating their programmes? Have they made any attempt to innovate participatory methods? What is the outcome? What are the reckons of the villages when PRA appraisals are conducted? What has been the course of its progress?

What are the problems in the application of PRA? What is the nature of such problems? What are the problems/challenges conquered by PRA practitioners? What are the problems which remain unsolved? These questions call for an empirical study. The study of this nature also gains importance as it would: (i) help document the state of art in PRA; (ii) facilitate improving the reliability of PRA; and (iii) help locate loopholes, if any, in the methods. Hence the study.

Review of Studies

Many studies have been conducted in the area of PRA. Most of the studies conducted earlier especially at the time of the advent of PRA are mostly related to application of PRA in different fields, strengths and shortcomings of PRA methods. Later, studies have been made either to test the validity of participatory methods by comparing them with other conventional methods or to establish the supremacy of participatory methods over the conventional methods. All such studies, conducted in 90's, are related to Third world countries. We review here a few of those studies.

Michael Richards, Jonathan Davies and William Cavendish[7] in their study "Can PRA Methods be Used to Collect Economic Data...?" have considered two main issues viz., (i) Can PRA produce similar economic data than traditional economic tools? and (ii) Are PRA techniques really more cost-effective especially when considering the time requirements of the participants? In order to find answer to these issues, the authors compared two main approaches viz., PRA and Survey by assessing the data sets for a range of variables including the production and economic returns of the main palm products—a variety of baskets and other craft products, and palm wine.

The findings of the study indicate that while fairly similar results were obtained for some variables, there were some major differences. The main area of discrepancy was in the production and sale of craft products. The PRA based estimates were unrealistically high when household budgets, based on PRA data were compared to normal household income in the area. The scoring and ranking of income sources also revealed an inconsistency in PRA data.[8]

For some quantitative aspects, the PRA exercise produced more reliable parameters, e.g., the labour requirements of the various craft products. The craftswoman PRA group very carefully deliberated the time required, going through each harvesting and processing stage and reaching a consensus. The survey did not have this flexibility; and enumerators reported that respondents experienced considerable difficulty with the concept of time (i.e hours). While this was also problematic for the PRA groups, it was possible to reach a common understanding through more extended discussion. PRA methods proved more effective in terms of differentiating and understanding the complete range of baskets and other craft products and were also better able to pick up temporal differentiation.[9] The survey proved more revealing in terms of gender differences.

The authors observed that time is one of the main factors in the analysis of cost-effectiveness. Researcher and facilitator time for the survey was about 60 hours, including the time spent on designing the forms, testing, enumerator training, and monitoring enumerators—about one and half a times that involved in preparing and facilitating the PRA exercises. But this narrow view of cost-effectiveness ignores the cost to the beneficiaries. It was estimated that community time spent in PRA exercises and community meetings was about 500 hours; about five times more than that spent on the household survey. Combining researcher/facilitator and community time reveals that the PRA absorbed approximately 240 per cent more of every one's time than the survey.[10]

The study revealed that each method has its particular strengths and weaknesses. The authors have, therefore, suggested complementary research methods, and pointed out for local and project purposes, a combination of participatory methods and small scale sample surveys is likely to be the most cost-effective means of determining the livelihood strategies of rural households. The two methods serve different and complementary purposes.[11]

Robert Davis[12] has tried to demonstrate how rapid appraisal and more traditional survey methods can be used together. The study carried out in Mauritania has its focus on risk management strategies that villagers use when faced with uncertainty. To achieve this aim, the author carried out RRA in six villages followed by a standardized survey, which was administered to 450 households in the same region. The RRA tools used include village maps, historical profiles, transects, consumption and historical matrices, Venn diagrams, wealth classification activities, and a series of semi-structured interviews on village crises, conflict, migration and development of village infrastructure.

The author has observed that use of RRA enabled the team to develop a more adequate and context specific research tool. The household based survey, on the other hand permitted the team to 'get-inside' households in a way that RRA could not.[13]

The unit of analysis in PRA, the author observed, is not the household but a kind of summary of typical behaviour of different subsections of the community. As such, the village or village sub-group was the unit of analysis. This was not always a limitation because certain exchanges occur between villagers and other outside entities. However, preliminary evaluation of survey results have shown that all the exchanges that were evaluated at the community level are used in varying frequency by individual households. The survey, thus, provided more specific insights into household behaviour and, by extrapolation, the frequency of various exchanges in the community.[14]

Taken together, the two approaches complemented each other. The great advantage of RRA phase of the study was its ability to deepen the description of risk management strategies and to evaluate change over time, things that no cross-sectional pre-coded survey can do. RRA gives voice to be people, in a way that no survey can.[15]

Sarah Gammage[16] has undertaken the research on PRA and its complementarities with household survey methodologies in El Salvador to estimate the returns to different mangrove management scenario and to compare the sustainable management of the mangroves to their conversion to shrimp and salt ponds. The study used both PRA and rural household survey for the purpose. Initially the study hoped to use PRA to describe the individuals and households relationship to resource and help develop a survey that could assess environmental inputs into the rural household economy.[17] However, as the PRA work evolved, the author felt that it could contribute more than just description of the eco-system and use of the environmental goods and services.[18]

The author observed further that during the course of data collection, PRA methods moved from being an additional tool that would furnish information about the community use of environment resources, to that of providing essential building blocks for the development of a community management strategy. What began as a focus group discussion, mapping and simulation exercise, gained momentum and integrated itself into the design and application of the rural household survey and ultimately informed a sustainable manpower management strategy.

The key PRA tools used in the study are: (i) separate focus groups and 'key informant' interviews with men and women to explore gender specific use of environmental inputs; (ii) mapping to define the boundaries of the village, identify the resources on which the village depended, identify the areas of mangrove deforestation by land use; (iii) calendar and time line to describe local agricultural, fishing, logging activities, external pressures on the local resource base and the loss of bio-diversity over time; (iv) various visual techniques and groups to generate information on the distribution of rainfall, water table fluctuation, and annual changes in fisheries resource. Transects were drawn of the estuary dealing depth at high and low tide, siltation areas, concentration of fish and shrimp throughout the year and boundaries of artificial shrimp of salt ponds.

The findings of the study revealed the limitations of quantitative methods like survey. Quantitative instruments can produce survey data and observations on variables that can be expressed numerically, statistically representative and easily extrapolated, although not always correctly.

Qualitative data produce extremely detailed, accurately location-specific socio-cultural information that may not be representative, may be subject to significant selectivity bias and cannot be extrapolated although not always correctly. However, in tandem, these methods yield rich picture of human activity. Furthermore, the strategic use of qualitative and quantitative methods can reduce the overall project cost and time burden of cumbersome and extensive quantitative instruments.[19]

One example shows these methodological complementarities. A survey was applied to 110 mangrove households and 489 individuals in El Tamavindo in 1993 and 1994. The survey took place in the wet and dry seasons to assess variations in fishing and agricultural activities. The survey revealed that 50 per cent of the men fished as their primary occupation while another 3 per cent were engaged in fish procuring and marketing. However, only one woman declared herself to be a fisher (person) and only 6 per cent stated that they were actively involved in fish processing and marketing.

The focus group discussion and key informant interviews, however, revealed a more complete picture. The qualitative data demonstrated that women do fish. While men fish in the open sea, the majority of the females confine their activities to the estuaries and shore line. Women are heavily involved in procuring the catches. They prepare dry fish for sale in the local and regional markets...

Using both the qualitative the quantitative data, the study team was able to determine that: 29 per cent of women regularly earn income from fishing and fish processing activities; approximately 60 per cent of all women clean fish and process the catch; 42 per cent clean boats and help their husbands moor and haul the catch in from the beach; and 17 per cent sell produce in local markets, restaurants and bars.

Narayanasamy, N., Mani, N.D. and ***Ramesh, R.***[20] in their study, mainly focussed on planning for Hunger Free Area Programme in a development block. Both survey and primary data were used. Three villages with total households of 157 were selected to have an indepth study of hunger and related issues. An important objective of the study was the identification of hungry families using participatory methods like mapping, wealth ranking, direct observation and livelihood analysis. The authors also made an attempt to crosscheck, reinforce and strengthen the information and reassure with the findings of participatory methods. Hence, a combination of three different conventional methods like diet survey, clinical assessment and anthropometric measurements were employed. The results of both the analyses were crosschecked and compared to see the validity of results obtained through PRA methods.

The findings of the study revealed (i) around 15 per cent of the 157 households have been classified as poorest of the poor based on locally evolved criteria; (ii) around 53 per cent are poor; and (iii) all the rest are classified as middle. The results obtained through nutritional survey, clinical assessment etc. have revealed that silent hunger in the form of malnutrition, under-nutrition and the resultant underweight are found in the poorest of the poor category. Juxtaposition of the people's classification of poor with the inference of nutritional status assessment has authenticated the objectivity in the criteria evolved by the people and also identification of poor/ultra poor families suffering from absolute/ silent hunger.

Rosemary McGee,[21] in her doctoral work in rural Columbia, seeks to explore the perception gap uncovering poor people's perspectives on their condition and to compare this 'bottom view' with 'top view' embodied in policy responses. The author has used a combination of research techniques, which include secondary data gathering, formal interviews at the national level, secondary data gathering and SSI at regional level and ethnographic research and rapid appraisal at village level.

The author, through her experience in the field, especially at the micro-level using RA and ethnographic research has concluded that RA for research on development purposes cannot be done rapidly and well in a totally unfamiliar territory. Without familiarity with the local context, it is likely to be ill-designed, ethically unsound and will generate local mistrust and spurious findings. Combination of the two methods have helped her to have a better basis for conducting methodologically and ethically sound RA that she had at the outset. This is partly because the ethnographic work is directly oriented to the design of RA.

Katherin Schreckenberg[22] discusses the experiences gained during her fieldwork in West Africa for a Ph.D. work. The author used a variety of methods to investigate the availability and use of indigenous non-timber products. These include conventional research methods such as ecological transects to measure tree density, fortnightly physiological observation of marked trees to examine seasonal availability of products, measurement of yields and regular market surveys. In addition, various PRA methods were employed ranging from matrices to calendars, maps and transects. The experience of the author reveals that: (i) application of conventional methods is fairly straight forward; can be applied in a very routine manner whereas PRA not only requires a researcher to have confidence in herself but more important to have confidence in the ability of local people to analyse and discuss; (ii) conventional methods of data gathering did not require much of adaptation to the local situation, whereas in contrast, although PRA methods require little adaptation to a specific situation their success depends on the facilitator having an intimate knowledge of the area in which they are being used; and (iii) all the conventional research methods were very labour-intensive for the researchers, but rarely used much of local people's time. In comparison, the PRA methods produced more results for a given unit of time, but required much heavier time investment from the participating villagers.

The author has found the combination of the two types of methods to be very rewarding, with PRA work supplying the contextual information for the quantitative data provided by the conventional research methods.

Michele Nori, Adane Hirpa and ***Giovanni A.Ferrar,***[23] *in* their case study, examine agro-pastoralism in Sub-Saharan Africa, exploring recent land use changes by analysing social and environmental processes. The authors aimed to explore land use changes in the research area by using a 'holistic approach' to soil science. This has been achieved by interweaving biophysical information with socio-economic and cultural data derived from local communities using PRA. The methods used are aerial photo, soil sampling and livelihood historical matrices.

The project area is characterized by physical diversity and uncertainty. If investigated only with classical methods and analysed according to objective parameters, it is likely that the area would be described as desertified. However, with a holistic approach it has been possible to understand endogenous development processes. This has led to a difference in the vision of the area itself, where the problems concerning social poverty and environmental degradation are tackled each day by local people with specific and often appropriate strategies.

The three types of investigation that the authors combined are complementary. Their integration enabled a reliable scaling up of information and enabled the authors to uncover a historical sequence of events. In particular, aerial photos showed great potential for promoting dialogue with local actors.

Atkinson,[24] in his study has used nutritional surveillance, as part of, or complementary to, the famine early warning system in Ethiopia, to collect reports on local food security from community leaders using structured interviews. It is important to assess the extent to which this information reflects the food-related behaviour of the community. Information on various socio-economic variables related to nutrition was collected at the household and community level through interviews in Western Shewa province. The data were compared and generally the correspondence between the two was good. Information topics which might be missed using only the local leaders, and ways to improve collection are discussed.

Andy Inglis,[25] in his study on "Harvesting Local Forests Knowledge ....", has compared the results of the study by using survey and participatory methods. First, a survey was conducted by a group of researchers in the area. This was followed by a PRA by a team of four persons. The survey was conducted by using a questionnaire consisting of 278 questions. The PRA techniques used by the team includes interviews, Key informant interviews, mapping, preference pair ranking, indicator identification, fuelwood measurement, and direct observation.

The study revealed that there are subtle variations in the results for some of the locations and a wide discrepancy for one of the major fuel wood using communities, Tombo. According to the questionnaire survey, Tombo has a very low level of species preference. The PRA survey found it to have the highest. Also, the most preferred species according to the questionnaire results did not appear in the top three species according to the special preference ranking of the PRA exercise. More alarmingly, this species, which was also top in a previous questionnaire survey (1986), was not mentioned at all in the 1990 formal survey results.

Again, from the results of the formal questionnaire surveys fuel wood marketing questions there are subtle variations in the results for all the locations and a wide discrepancy for one of the locations, Songo. The PRA survey for Songo clearly shows a well-organised fuel wood trading system satisfying both local and external (urban) demands, operated by eight local static traders. However, the tabulated questionnaire results indicate that there are no fuel wood sellers at all in Songo! This is despite the same questionnaire survey producing a result, which indicates that 75 per cent of the community buy fuel wood.

As a contribution to the PRA versus formal statistical survey debate, the comparison between the results obtained by the different methods throws up interesting similarities and differences. It could be argued that, backed up by the available evidence, the two main variations described above indicate that PRA is a far more appropriate and accurate way to collect socio-economic and socio-ecological information for social forestry projects.

Gill[26] has compared seasonality diagrams drawn up by farmers during an RRA/PRA workshop carried out in Western Nepal, with scientifically collected meteorological data from a nearby agricultural research centre. The paper considers a number of methodological issues connected with the linking between formal and informal methods of research. It then moved on to consider the question of fitness between the two sets of data. It concludes that if the scientific data is the real data, then in fact PRA data is extremely accurate. In terms of cost effectiveness, the PRA data is better, as it was collected extremely quickly and effectively without the need for an expensive infrastructure such as the agriculture research centre.

Lindberg, C., Loiske, V.M., Ostberg, W.N., and ***Mungongo, C.***[27] present a critique of an agency study, which used rapid research methods to investigate the role and consequences of structural adjustment programmes and the introduction of multi-party system in Tanzania. The authors compare the findings of the agency study to their own village-level studies. They argue that the genuinely poor were not included in the analysis by the agency study team, thereby casting doubt on the study's provisional findings that 'trade liberalization has been good for rural people'. They suggest that special efforts need to be made to ensure that 'the unseen and unknown come to the fore' when using rapid research methods.

Mosse, D.[28] in his article explores constraints encountered when using PRA on an ODA-funded natural resource project in a tribal area of western India. It was particularly evident that women participation in the PRA was minimal. The reason for this were practical-(Women were not available collectively for long period of times and there were few women field workers as the project had just begun) social-(PRA activities tended to take place in public places where women felt awkward) and methodological (women respond to PRA activities in different ways, sometimes feeling bored and communicating by singing instead). The author argues that an organized PRA gives privilege to certain kinds of knowledge and representation and suppresses others. The emphasis given to formal knowledge and activities tends to "reinforce the invisibility of women's role". However, once the formal and public nature of PRA is perceived as a problem, it can become a means by which women's knowledge and activities can be transferred from the informal to the formal arena of project planing.

Natarajan, I. et al.,[29] evaluated the National Programme on Improved Chulah (National Council of Applied Economic Research: NCAER) to examine the efficacy of complementing or substituting sample surveys (SS) with RRA/PRA, based on comparison of results, reliability level of data sets, cost/time trade off, quality of field workers, and training components. The comparisons were carried out only in the Maharashtra area and attempts were made to maximize the similarities in sampling procedures, places and time of field work, so that a proper comparative assessment of the two approaches could be facilitated. The findings are as follows :

— at the village level, comparison of quantitative as well as qualitative data clearly point to the higher accuracy and reliability levels of the data emerging from the RRA/PRA methods as compared to the village schedule based data in SS.

— at the regional/state level, the feasibility of employing PRA methods in generating quantitative data for evaluation must be questioned, since the need to increase the number of village based RRA/PRA surveys will significantly raise costs (the data collection costs per village in RRA/PRA based studies was seven times more than that of the SS based study).

— on the whole, the blending of the two questions appears to be the logical and feasible propositions. However, this will call for a substantial modification in procedures for organization of field work, sample selection, recruitment and training of field workers. The report emphasizes the differences in evaluator-investigator training procedures that characterize the two approaches: the RRA/PRA training process has a significant potential (and actual) impact on improving the quality of the evaluation findings while it is observed SS training has little impact on improving the reliability of the collected data.

Scoones, I.[30] compares the results of wealth ranking and household survey approaches, which were applied in tandem for a sample of farming households in southern Zimbabwe: It concludes that wealth ranking provides an accurate indicator of relative wealth and that ranking can be a useful complementary methods to be employed alongside survey assessments. Qualitative discussion during ranking exercises reveals details of the historically, socially and economically constructed understandings of wealth and well being of different actors. The conventional assumption that surveys always provide better data is thus questioned.

Neela Mukherjee[31] in her study on Participatory Rural Appraisal and questionnaire survey, has compared conventional methods of investigation like questionnaire survey with participatory methods like PRA-type enquiries as evolving in recent years. She has largely relied on field reports which related to experience involving relatively large areas and reflect parallel use of questionnaire survey and PRA type inquiry or some combination of both. She has generally drawn on :

— those studies which have used either or both conventional survey and PRA-type method for data collection;
— those which show scale effect in use of PRA type method; and
— those which use a combination of the two kinds of inquiry for data collection.

The study has proved that data collection whether through questionnaire survey or PRA method involves basic issues in cost time, scale, utility, depth and other aspects. In addition, it is important to achieve a reasonable level of trustworthiness. The study revealed that it is possible to use PRA type inquiry to conduct large-scale data collection and get quicker results with deeper analysis. The author has concluded that for any type of field inquiry, it is best to approach it with an open mind and explore a few questions in order to know more as to why an inquiry is being conducted, what are its objectives, what is the output required, in what form, for whom and when. All these questions can then be weighed for considerations such as time, cost, quality and quantity, trustworthiness of data and the like. As such there are no simple answers existing to guide field level enquiry since the requirements, objectives, constraints, options and conditions operating for a field enquiry differ from situation to situation. In each case, it is best to take a pragmatic assessment of the situation at hand and deliver meaningful results whether through exclusive application or combination of the methods at hand.[32]

Rajaratnam et al.,[33] have validated the methods of wealth ranking by PRA vs survey in identifying the rural poor. The objectives of the study are : (i) to identify the type of indicators used by the community to classify their economic status; (ii) to validate professionals' classification with community classification using wealth ranking technique of PRA; and (iii) to determine the reasons for wrong classification of certain households by community and by professionals. The study was conducted in three different centres covering 412 households by using both questionnaire and PRA method. The steps followed in the study were: (i) prioritization of households initially by professionals and by community; (ii) matching of prioritization of the two groups (iii) verifying the different priorities by both the methods under consideration; and (iv) validation of prioritization along with reasons for prioritization.

The results of both the survey and wealth ranking were matched using comparison and the percentage of households with same and directed priorities were also calculated. It was found that on an average, only 62 per cent of the professional priorities for identifying households coincided and matched with the priorities set by the community. This made for a difference of 38 per cent in the two priorities. An in-depth investigation was taken up for 50 per cent of the 38 per cent of those households which fell into the category of different priorities of professional and of community. Such further probing of various key informants and the households concerned and cross-checking them with PRA and survey results helped the investigators to finalise the actual economic level of the households in question. On crosschecking of data the study concluded that for the area under consideration, the professionals through their fixed indicators were, on an average, 57 per cent accurate in identifying the economic levels of the household concerned, whereas, the community using the wealth ranking technique was, on an average, 97 per cent accurate in identifying the economic levels of the households. The study also has documented the reasons for wrong classification by the professional which include concealment of information by the community on ownership of land, livestock, assets, income sources and actual income, (i) investigators' failure to get information on subsidiary occupation and (ii) fixed indicator of professionals.

National Council of Applied Economic Research[34] made an attempt at using RRA/PRA method to focus on the rural dairy sector in Operation Flood areas of selected regions of India. It had earlier done a cross-sectional sample survey in all operation Flood areas, which had generated valuable data set. The RRA/PRA exercises were undertaken in some of the villages in the three regions studied earlier. They were topical PRA exercise on dairy cooperation and milk economy of the regions under study.

The results of the RRA/PRA were close to those of the earlier survey on some of the major parameters. The RRA/PRA study generated additional information, which confirmed some of the linkages and behavioural patterns. The study has concluded that RRA/PRA approach cannot substitute for sample surveys. However, the former effectively complement and supplement the latter. RRA/PRA can be integrated with the conventional surveys in such a manner that the quantitative/estimation aspects are left to sample survey and qualitative aspects are confined to RRA/PRA. The study team was of the opinion that RRA/PRA method can be effective in guiding pilot study for proper planning and implementation of the main survey.

Buchanan Smith, et al.,[35] as part of the FAO assistance for the development of National Food Security System in Chad, attempted an informal survey method using PRA techniques to try and understand how people perceive and prioritize their food security problems. The survey design and methodology for the exercise drew on the experience of food security surveys carried out at a sub-sectional level in Darfur and Sudan. The survey was conducted between June and September 1991.

Three concepts formed the food security planning approach: (i) the availability of staple food; (ii) the stability of supply of staple food; and (iii) access of consumers to staple food. The survey had the objective of finding out how people at the grassroots level, perceive their food scarcity problem in their daily lives and solutions they propose. The final output of each of these national programming exercises is intended to be a range of alternative, strategies to achieve food security, from which national government can select its own policy options.

The reasons for choosing an informal survey are: (i) the survey was to be participatory; (ii) it was to be carried out within a short period and with limited resources; and (iii) the results were to be mainly in qualitative form. There were 14 prefectures in Chad, all to be included in the survey. Within each prefecture, the survey was designed according to the principal production systems to ensure some generalizations of the food security problems and solutions raised. The process at the local level started with the help of local authorities on the basis of which a representative village was selected for each production system. The survey was conducted in each of such village, which were 55 in number. The survey team spent on an average, one day in each village and the techniques of semi-structured interview was adopted, first with a group on the basis of the checklist of questions to ensure some degree of standardization. This was followed by interviews at the household level to crosscheck the information gathered in groups. At the household level, the women were mostly talked to with the help of female enumerators. The survey was also conducted in large number of towns, where different urban groups were identified and interviewed individually as per checklist. This was followed by ranking exercise of food security problems and suggested solutions.

The enumerators prepared case studies of village on the basis of a standardized format by assimilating and organizing all the qualitative information collected during the interview. The final analysis involved assimilation and organization of materials from all such case studies in a form useful and coherent for national planning. To facilitate such a task, the results were discussed with the enumerators so as not to lose any relevant knowledge or experience. The huge and unwieldy data was made systematic and reduced by using techniques to summarize information concisely and sometime visually. Frequency tables of problems and solutions were summarized and averages made. The background information on rural economy, recent changes, food security system and coping strategies were documented.

The authors have pointed out that with limited time and availability of resources, RRA/PRA approach permitted a survey of this scale to be carried out with very wide geographical coverage. Only this kind of survey approach could have enabled people's views to be incorporated into a national level planning process. This represents a very important and quite unusual attempt to consult local people when formulating a national programme. The survey has provided concrete suggestions to discussions that have been going on in the national capital about strategies to promote food security, which stuck at a rather general level and lacked the necessary details to be translated into action.

Among studies reviewed many have proved that combination of PRA and conventional methods would be rewarding. However, these studies have not tried to capture the views of NGOs and the opinion of people at the grassroots on the efficacy of participatory methods. The current study with its focus on studying the state of art of PRA among NGOs in Tamil Nadu, the problems and challenges in practising participatory methods would throw more light on the participatory methods with emphasis on their efficacy.

Scope of the Study

The study has been undertaken with the objective of studying the state-of-art of PRA among NGOs in Tamil Nadu with a primary focus on efficacy of participatory methods. Participatory methods have been introduced in development work as they are found to be participatory, enabling and empowering the primary stakeholders of the project viz., the people. The continued practice, primacy and supremacy of these methods depends on its efficacy in the field. The study mainly aims at ascertaining the efficacy of participatory method as experienced by the NGOs and also the people. Closely related to efficacy are problems and challenges faced by the NGOs in participatory methods. The study therefore has its focus on these areas.

Objectives

The objectives of the study are:

(i) to assess the state of the art of PRA, prevailing among NGOs in particular;
(ii) to find out the problems/challenges encountered in employing PRA methodology in rural development work by NGOs;
(iii) to understand the steps taken to overcome such problems and challenges and the effectiveness of such steps;

(iv) to ascertain the efficacy of participatory methods; and
(v) to ascertain the innovations, if any attempted by NGOs in the arena of participatory methods.

Hypotheses

The study had the following as its hypotheses :

- Use of PRA has brought about a near elimination of the feelings of hierarchy among the staff.
- The information collected through PRA is more reliable than the one collected through survey.
- Domination and vested interest of a few in PRA affects the quality and reliability of the information.
- Failing to constitute a good representative group of people from the village with heterogeneous nature affects the quality of the PRA outcome.
- Narrowing down of social distance and empowering people through PRA process are all temporary illusions and they do not last long.
- Empowerment is possible only when PRA process leads to institution building.

Operationalization of the Concepts

Participatory Methods: They are the methods used under Participatory Rural Appraisal for generating information, planning, implementation, monitoring and evaluation of rural development activities.

Efficacy: Efficacy means the effectiveness of the participatory methods. Efficacy of participatory methods was assessed in terms of reliability, timeliness, rapidity, cost-effectiveness, simple to analyse, easy to practice, accuracy, appropriateness, participation, adequacy, sharing of information, creation of positive attitude and behaviour.

NGOs: Non-Governmental Organisations include a wide variety of groups. Definitions vary, but NGOs typically are entities that are entirely or largely independent of government and have humanitarian or cooperative rather than commerical objectives. For the purpose of our study, we define NGOs as private organisations that pursue activities to relieve suffering, promote the interests of the poor, protect the environment, provide basic social services, or undertake community development.

Methods are defined as comprehensive set of approaches to gather, evidence and analyse the specific problems, encompass techniques and tools.

Techniques are designated as practical ways of collecting data and for analysing the information obtained in the research process.

Tools are the instruments that are used in the process.

Methodology

This is an empirical study based on a mix of methods, techniques and tools. First hand data, both qualitative and quantitative, were collected from the heads, project coordinators, field level workers of the selected NGOs and also from the people mostly through semi-structured interviews. Questionnaire survey, case study method, and participatory methods were used.

Unit of study: The study is confined to Non-governmental organisations practicing PRA in Tamil Nadu.

Selection of NGOs

A list of NGOs practicing PRA in Tamil Nadu was obtained from Tamil Nadu Resource Team* (TNRT). There were 72 NGOs enlisted with TNRT. A questionnaire was sent to all the 72 NGOs; of which 54 responded. All the 54 NGOs come under the purview of this study. We selected four NGOs** for an indepth analysis by using Case study method. The NGOs selected are: (i) OUTREACH, Bangalore; (ii) SPEECH, Madurai; (iii)ASSEFA, Lathur; and (iv) Tamil Nadu Women's Collective, Chennai. The selection of these NGOs is mainly based on age, experience, practice of participatory methods at different stages of the project, their commitment and conviction to participatory methods. They have common features :

---

* TNRT is a network organisation for the NGOs which practice PRA in Tamil Nadu.

** Totally five NGOs were considered in order to study the efficacy of participatory methods with reference to select variables as perceived by NGOs. They are: i) OUTREACH, Bangalore; ii) SPEECH, Madurai; iii) ASSEFA, Lathur; iv) SEED Trust, Nagercoil, and v) RENAISSANCE, Pudukottai.

- — they are the pioneers in applying PRA techniques in their rural development activities;
- — all the staff including the chief executive are trained in PRA;
- — they are the first generation practitioners of PRA;
- — they have made participatory methods as part of their work culture; and
- — they have integrated participatory methods in their project.

These qualities were assessed based on focussed group discussion with the NGOs which responded to our mailed questionnaire.

Tools for Data Collection

Bearing in mind the objectives of the study, a combination of tools were employed in order to gather information and data.

To begin with, a short, mailed questionnaire was used to understand the genesis of NGOs, their objectives, their activities, the range of experience in PRA and problems in organising PRA in villages. This was followed by focussed group discussion with project coordinators, executives and field workers of selected NGOs who were involved in organising and conducting PRA in their service villages. The focus group discussion based on semi-structured interview technique canvassed information on the current status of PRA in each organisation, the problems in using PRA techniques, ways and means of overcoming the problems, the strengths, the weaknesses, opportunities of PRA and threats to PRA techniques. A group discussion with the NGO functionaries was arranged to evolve criteria for ascertaining the efficacy of participatory methods. Scoring technique was used to measure the efficacy of each criterion evolved by the participants. Observation method was used to see how participatory training programmes were organised, how the meetings of self-help groups were convened, and how participatory methods were used in their appraisal, analysis and planning.

In order to compare the participatory methods with the other conventional method like survey, a survey using an interview schedule was conducted in three villages. Participatory mapping was done in the same three villages to compare the results obtained from the two methods.

The villagers - men, women and youth who participated in PRA programme (villages coming under the jurisdiction of selected NGOs) organised by NGOs were met in groups, to elicit their views on PRA methods, its strong and weak points.

Framework of Analysis

Data gathered are largely qualitative in nature. Descriptive statistics were used to analyse the data. Averages and percentages were used to present the profile of the NGOs with reference to their evolution, jurisdiction, objectives, sector of activities, clients/ beneficiaries, method of operations, problems faced and also to analyse the state-of-art of PRA among the NGOs with regard to experience in PRA, source of training, PRA methods used, practical application, benefits of using PRA, benefits to people, problems in practicing PRA, solution to problems, etc. To study the efficacy of participatory methods, participatory methods as practiced by four selected NGOs have been analysed. Inferences from such analysis were drawn. Further, criteria for assessing the efficacy of participatory methods in consultation with NGOs were evolved. Then they were asked to assign scores for each criterion out of 10 scores for each criterion. The scores were then averaged and analysis made. The efficacy was further tested by using the results of the survey conducted in three villages with that of results obtained out of participatory mapping.

Report Structure

The report has six chapters. The first chapter provides an overview of the Participatory Rural Appraisal with particular reference to its genesis. Chapter 2 reports on the design of the study. Chapter 3 contains the profile of NGOs covered under the study and practice of PRA by NGOs. Chapter 4 is an indepth analysis of participatory methods as practiced by NGOs. Chapter 5 reports on efficacy of participatory methods as perceived by selected NGOs and also as ascertained at field level. Chapter 6 contains findings of the study and conclusion.

NOTES AND REFERENCES

1. Mukherjee, Neela., *Participatory Rural Appraisal: Methodology and Applications*, Concept Publishing Company, New Delhi, p. 26.
2. Chambers, Robert., "Rapid Rural Appraisal–Rationale", IDS, *Discussion Paper No.155,* IDS, Sussex, 1981.
3. Conway, G.R., "What is in a name?", Editorial, *RRA Notes,* No.1, June 1988, p. 3.
4. Mukherjee Neela, *op.cit.*, p. 29.
5. PRA Unit, Gandhigram Rural Institute, *PRA Retreat in a Rural Setting Gandhigram,* Gandhigram, Tamil Nadu, 1995.
6. Jules N. Pretty., *Regenerating Agriculture: Policies and Practice for Sustainability and Self-Reliance*, Earth Scan Publication Ltd., London, 1995, p. 178.
7. Michael Richards, Jonathan Davies and William Cavendish, "Can PRA Methods be Used to Collect Economic Data? : A Non-Timber Forest Product Case Study from Zimbabwe", *PLA Notes 36*, October 1999, pp. 35-40.
8. *Ibid* P. 35.
9. *Ibid* p. 37.
10. *Ibid* p. 38.
11. *Ibid* p. 39.
12. Robert Davis, "Combining Rapid Appraisal with Quantitative Methods: An Example from Mauritania", *PLA Notes* 28, Feb., 1997, pp. 33-36.
13. *Ibid* p. 35.
14 *Ibid* p. 35.
15. *Ibid* p. 36.
16 Sarah Gammage, "PRA and its Complementarities with Household Survey Methodologies", *PLA Notes 28*, Feb.1997, p. 47-54.
17. *Ibid* p. 47.
18. *Ibid* p. 47.
19. *Ibid* p. 48.
20 PRA Unit, Gandhigram Rural Institute, *Hunger Free Area Programmes: A Pilot Study*, Gandhigram, Tamil Nadu, 1997.
21. Rosemary McGee, "Ethnography and Rapid Appraisal in Doctoral Research on Poverty", *PLA Notes 28*, Feb.1997, pp. 55-58.
22. Katherin Schreckenberg, "The Respective Merits of PRA and Conventional Methods for Long-Term Research", *PLA Notes 24,* October 1995, pp.74-77.
23. Michele Nori, Adane Hirpa and Giovanni A. Ferrai, "Complementary Methods to Understand Land Use Changes: An Example From the Ethiopian Rifts Valley", *PLA Notes 35,* June 1999, pp.16-20.
24. Atkinson, S. "Interviews with Key Informants and Household Surveys: Central Ethiopia", *Disaster,* Vol.16 No.3, pp. 255-258.
25. Andy Inglis, "Harvesting Local Forests Knowledge: A Comparison of PRA and Conventional Surveys", *RRA Notes 12*, July 1991, pp. 32-40.

26. Gill. G., "But how does it compare with real data", *RRA Notes 14,* Dec. 1991, pp. 5-13.
27. Lindberg.C., Laiske, V.M., Ostberg. N., and Mungiongo, C., "Handle with Care! Rapid Studies of the Poor", *PLA Notes 22*, Feb. 1995, pp. 11-16.
28. Mosse. D, Authority, "Gender and Knowledge: Theoretical Reflection on PRA", *Economic and Political Weekly* 30(11), March 18, 1995, pp. 569-578.
29. Natarajan. I., et.al., "Comparative Study of Sample Survey and PRA Methodologies with special reference to Evaluation of National Programme on Improved Chulah", NCAER, New Delhi, 1993.
30. Scoones.I, "Investigating Difference: Application of Wealth Ranking and Household Survey Approaches among Farming Households in Southern Zimbabwe", *Development and Change*, Vol. 26, 1995, pp. 67-68.
31. Neela Mukherjee, *Participatory Rural Appraisal and Questionnaire Survey,* Concept Publishing Company, New Delhi, 1997.
32. *Ibid.,* p.137.
33. Rajaratnam, Jolly, Ganeshan, C., Tharian, Helen, Babu and Abel., "Validating Wealth Ranking of PRA and Formal Survey in Identifying the Rural Poor", RUSHA Department, Christian Medical College & Hospital, RUSHA Campus, Vellore (TN), April 1992.
34. National Council of Applied Economic Research, *Evaluating the Impact of commercialization of Rural Dairy Sector: An application of RRA Techniques*, New Delhi, 1992.
35. Margaret Buchanan Smith, *et.al*., "Finding out how people prioritize their Food Security Problem in Chad: The Challenges of PRA at National level", *RRA Notes* 18, June 1993, pp. 33-43.

3

STATE OF ART OF PRA AMONG NGOs IN TAMIL NADU

PRA which initially sounded like a piece of certain theoretical propositions, has turned out to be in fact a scientifically conceived concept based on continuous field trials, tailored to socio-geographic reality and its requirements. It has been slowly taking roots in the soil of villages as well as the organizations of social scientists, with wider ramifications and deeper implications for project formulations, resource appraisal, manpower planning, information generation and allied tasks in the field of rural development.

In the process of contributing to growth, PRA, by itself is undergoing a metamorphosis, as it is used by a multitude of agencies from villagers to World Bank officials. From merely the NGOs, it is being handled by a host of other agencies such as Educational Institutions, Research centres, Training Institutes, Trade Unions, Trusts, Government Departments, Gandhian Institutions and Grassroots organizations.

PRA is passing down from social scientists to students, scientists to social workers and researchers to ryots. What was once a luxury, PRA is fast becoming a necessity to planners, policy makers, teachers, students, social workers, scientists and scores of others. From its erstwhile status of having to be 'sold' from the experts to the clients, it is being seriously 'sought after' either for greater specialization in the area of work or at times for sheer survival in some of the change agencies.

By mid-1996, PRA picked up in perhaps 100 countries and there were over 30, mostly national PRA related networks. Application had become legion in almost every sector of field-level development[1]. The applications of PRA in main sectors include Natural Resources Management, Agriculture, People, Poverty and Livelihood, Health and Nutrition and Gender. Other areas where PRA is applied include Adult Literacy, Children's Education, Emergencies and Refugees, Organizational Analysis, Participatory Monitoring and Evaluation, Sexual and Reproductive Health and so on.

Tamil Nadu, through its NGOs, is a pioneer in spreading the concept of PRA to every nook and corner of the State. NGOs like SPEECH (Madurai), ASA (Trichy), SEVAI (Kulithalai), and Women's Collective (Chennai) have been mainly responsible for early and quick spread of participatory techniques. They have also discovered and rediscovered certain participatory techniques and made rich contribution to the existing menu of PRA methods. Among the educational institutions, Tamil Nadu Agricultural University (Coimbatore) and Gandhigram Rural Institute have done commendable work in taking PRA to other universities, research and training institutions by organizing training programmes, conducting research and offering consultancy services.[2]

The quick spread of PRA and demand for training in PRA have necessitated a sort of networking among the practitioners of PRA which resulted in the birth of Tamil Nadu Resource Team (TNRT) in the year 1992. The primary aim of the network (TNRT) is to keep up the quality of PRA, offer field based participatory training in PRA, innovate new methods in the field of PRA, provide a platform for sharing the field based learning experiences, act as a clearing house and so on. There are about 72 NGOs in the network. They employ PRA techniques in their organization. Adoption varies from appraisal of the existing situation to analysis of their organizational problems. Some NGOs like SPEECH, Women's Collective and ASA have made PRA as part and parcel of their activity and have been continuously using it. Several other NGOs do practice PRA at different levels in a variety of fields.

We present, in this chapter, the state of art of PRA in Tamil Nadu with special reference to the experience in PRA, training undergone, application of PRA, benefits derived in using PRA, problems encountered in employing PRA, measures taken to overcome such problems and other related issues. A profile of the NGOs has been sketched as a prelude to the analysis. We present here the profile of the selected NGOs.

PROFILE OF NGOs

NGOs—their Evolution

NGOs have been functioning in Tamil Nadu for quite a long period of time. Our country's independence gave an impetus to voluntary work. Many NGOs were initiated on Gandhian lines after independence mainly on the advice of Gandhiji to his constructive workers to go and serve in rural areas based on his 18-point Rural Reconstruction Programme. The Christian Missionaries too were mainly responsible for the quick growth of NGOs as they started working in the areas of health and education. Recognizing the efficacy of NGOs in rural development work, the government encouraged the NGOs by providing financial support in their development work. This has provided tremendous fillip to the growth of the existing NGOs and birth of new NGOs. The spurt in the growth of NGOs could be seen in 1980s.

Response of the NGOs

Of 72 NGOs under the fold of TNRT, 54 responded to our questionnaire. The profile of NGOs are presented in terms of their years of experience, jurisdiction, objectives, activities, their clients, methods followed at different stages of the project, problems faced in different stages of the project and the steps taken to overcome such problems.

Years of Service

Majority of the NGOs, as the table shows, have been started in

80s and 90s. Only 9 per cent of the NGOs were established before 1980. The service period of the NGOs shows that around 85 per cent had sufficiently long years of experience. Only 15 per cent of them are young with less than 5 years of experience. (See Table 3.1)

Table 3.1: Years of Service of the NGOs

Age in Years	*Number*	*Percentage*
< 5	8	14.8
5 – 10	14	25.9
10 –15	16	29.6
15 – 20	11	20.4
20 and above	5	9.3
Total	54	100.0

Objectives

Objectives are the policy statements. They largely decide the activities of the organisation. The objectives of NGOs could be found in the bye-laws or the rules and regulations that govern the working of the NGOs. A perusal of the byelaws of NGOs normally indicate listing of a large number of objectives touching almost every aspect of rural development that one can imagine. Many of them may remain in the book without being realized. Hence, they were asked to present the most important and pursued objectives of their organization. Their responses are consolidated and presented in Table 3.3.

The most important objectives of the organizations, as reported by the NGOs, are arranged below :

— awareness creation on social, economic and political issues among different sections of the community (87.0%);

— empowerment of women through awareness creation, formation of self help groups and promotion of income generation programme (87.3%);

— undertaking socio-economic welfare activities among children, women, the aged and handicapped (85.2%);

— promotion of credit and savings (85.2%);
— eradication of illiteracy, ignorance, economic dependency and exploitation (74.1%);
— strengthening of people's organisation (68.5%); and
— empowerment of the poor and oppressed (55.6%).

Certain objectives as can be seen overlap. For instance, promotion of credit and savings is closely associated with empowerment of women through self-help groups. Less prominent objectives in terms of number of responses are:

— networking of self-help groups (9.4%);
— popularizing organic farming through vermiculture (9.4%);
— prevention of female infanticide (11.1%);
— development and popularization of alternative technology (13.0%); and
— entrepreneurship development (13.0%).

The objectives of the organizations indicate that a vast majority of them work among the women and other marginalised sections of the community with a view to empower them through awareness creation, organising, initiating income generation and other socio-economic programmes. (See Table 3.2)

Sector of Activities

All the organizations are working in more than one sector. Table 3.3 shows the sectors and different activities undertaken in each sector. The main sectors of work as reported by the respondents are women and development, income generation programmes, livestock, health and sanitation, agriculture, forestry, education, child development, housing, ecology, protection of human rights, etc. The activities undertaken in different sectors of work show that the NGOs mainly concentrate on awareness, skill development and securing livelihood opportunities, which would result in increased income to the families. This was followed by provision of safe drinking water, better environment, promotion of health care, education of different type viz., pre-school education, supplementary education, non-formal education etc.

Table 3.2 : Objectives of the NGOs

Objectives	*Number of responses*	*Percentage n=54*
1	2	3
Empowerment of women through awareness creation, formation of self-help groups, promotion of income- generation programmes	47	87.0
Awareness creation on social, economic and political issues among different sections of the community	47	87.0
Promotion of credit and savings	46	85.2
Undertaking socio-economic, welfare activities among children, women, the aged and handicapped	46	85.2
Eradication of illiteracy, ignorance, economic dependency and exploitation	40	74.1
Strengthening of people's organization	37	68.5
Empowerment of the poor and the oppressed	30	55.6
Facilitating the marginalized, especially the dalit and the women, to organize themselves through awareness education and training	23	42.6
Integrated rural development	20	37.0
Undertaking activities to better the health conditions of sex workers, slum women, SC women and truckers	20	37.0
Support to pre-school, primary and non-formal education in rural areas	17	31.5
Assisting the community in prevention and control of communicable diseases	15	27.8
Protection of human rights through advocacy and legal reforms	15	27.8
Promoting the socio-economic conditions of the people in drought prone-areas	15	27.8

Contd......

Table 3.2 *(Contd.)*

1	2	3
Provision of health facilities	15	27.8
Conservation of natural resources	12	22.2
Support to agricultural activities among the small and marginal farmers	12	22.2
Creation of safe environment	11	20.4
Elimination of child labour	10	18.5
Development and propagation of Alternative Technology	7	13.0
Promotion of self-reliance through entreprenuership development	7	13.0
Prevention of female infanticide	6	11.1
Popularising organic farming through vermiculture	5	9.4
Networking of self-help groups	5	9.4

A comparison of the objectives of the NGOs and the activities undertaken by them indicates a close relationship between these two.

Jurisdiction of NGOs

NGOs, largely confine themselves to limited areas of operation. They are, by and large, small. Their small size with limited jurisdiction offer them several advantages like close cooperation with the clients, a better understanding of the local problems, familiarity with different sections of the community, face-to-face interaction, continuous and consistent touch with the community, adjusting themselves to the local situation, and better staff relationship.

All these factors help them to plan better and work better with active support, cooperation and participation of the people. It helps them to "work with" the people. The results are also, invariably, sustainable. On the other hand, when an organization grows and becomes bigger, the advantages of 'smallness' disappear. The

Table 3.3: Sector-wise Activities Undertaken by NGOs

Sector of work	*Activities*	*Number of responses*	*Percentage n=54*
1	2	3	4
Agriculture	Land development activities	24	44.4
	Supply of inputs	20	37.0
	Watershed management	10	18.5
	Tank rehabilitation	6	11.1
	Training in organic farming	5	9.3
	Vermicomposting through women groups	5	9.3
	Sericulture	5	9.3
	Watershed resources center	2	3.7
Forestry	Tree planting	20	37.0
	Training in nursery raising	15	27.9
	Agro-forestry	10	18.5
	Tank-bund forestry	9	16.7
Livestock	Promotion of dairy activities	45	83.3
	Veterinary health centre	5	9.3
Water and Sanitation	Awareness programme on sanitation	45	83.3
	Provision of safe drinking water	36	66.7
	Construction of toilets	30	55.6
Education	Supplementary education (Tuition centre)	22	40.7
	Pre-school education	20	37.0
	Adult education	17	31.5
	Non-formal education centre	11	20.4
Health	HIV/AIDS awareness	35	64.8
	Health centers	7	13.0
	Leprosy & TB control	5	9.3
	Community health development	3	5.6
	Herbal treatment	2	3.7
Income Generation	Income generation activities	51	94.4
	Skill training	48	88.9
	Rural technology centres	2	3.7

Contd......

Table 3.3: *(Contd.)*

1	2	3	4
Women Development	Leadership training	52	96.3
	Campaigning for women's rights	52	96.3
	Organising women	50	92.6
	Formation of self-help groups	47	87.0
	Starting of micro-enterprises among women groups	40	74.1
	Strengthening the participation of women in panchayats	20	37.0
Child Development	Eradication of child labour	10	18.5
	Child labour action research	5	9.3
	School for mentally retarded children	1	1.9
Housing	Construction of houses	20	37.0
	Repair of houses	10	18.5
Ecology	Awareness programme	25	46.3
	Training programme in preservation of ecology	20	37.0
	Campaign against loss of ecology	15	27.8
Human Rights	Protection and promotion of human rights of women, children, dalits and unorganised workers	15	27.8
	Monitoring human rights protection	13	24.1
	Documentation and dissemination of human rights related issues	7	13.0
Others	Organizing the youth	39	72.2
	Legal assistance	15	27.8
	Legal literacy	15	27.8
	Dissemination of information	10	18.5
	Training in PRA	5	9.3
	Networking of women groups	5	9.3
	Networking of SHGs	5	9.3
	Community banking	3	5.6

organization then becomes 'formal', 'bureaucratic', and tends to be 'secluded' from the people. But then 'small' and 'big' are relative terms.

What should be the size of an NGO? There is no yardstick to measure the size of the NGO. However, it can be said that two or three areas encompassing revenue villages with 20 to 30 hamlets might be an optimum size to keep all the good qualities of an NGO in tact. Going by this standard, it can be said that many of the NGOs under report are comparatively bigger.

More than 40 per cent of the NGOs have a jurisdiction of more than 50 villages. Around 29 per cent of the NGOs can be said to have smaller jurisdiction while another 28 per cent may be said to have medium-sized jurisdiction. (See Table 3.4)

Table 3.4 : Jurisdiction of Non-Government Organisations

Number of villages	*Number of NGOs*	*Percentage*
Less than 10	4	7.4
11-20	10	18.5
21-30	2	3.7
31-40	5	9.3
41-50	10	18.5
Above 50	23	42.6
Total	54	100.0

Clients/Beneficiaries

NGOs normally work among the economically poor and socially marginalised sections. NGOs under report are no exception to this. Their clients include women, children, youth, scheduled caste, labourers, etc. They also serve the interests of institutions like Government departments and people's organization like local bodies. NGOs under report do not concentrate on only one group but many groups. The dominant groups as reported by 96 per cent of the NGOs are women; labourers (78 %); scheduled caste (72 %); others (65 %). Thus, the concentration of NGOs seems to be on women, scheduled caste, labourers, and others (See Table 3.5).

Table 3.5: Beneficiaries/Clients of the NGOs

(n=54)

Benefici-aries	*Nature & Type*	*No. of NGOs*	*Percentage*
Women	Poor, deserted & destitute, unemployed rural women, SC women, sex workers, slum women	52	96.3
Labourers	Landless labourers, dalits, unorganized	42	77.8
Dalits	All sections	39	72.2
Others	NGO, Government, backward communities, victims of human rights violation, truckers, people's organizations	35	64.8
Children	Poor, child labourers	29	53.7
Farmers	Poor, marginal, small, large	24	44.4
Youth	Poor, unemployed, dalit, students	22	40.7

Method of Operation

NGOs work with the people through a project. They assess the situation at the ground level, ascertain their problems, formulate projects, submit the proposal, obtain funds, implement, monitor and evaluate the projects. The methods followed at different stages of the project decide and determine the realization of objectives of the project. Three major phases of the project are formulation, implementation and evaluation. What types of methods/tools/techniques are adopted by the NGOs? Table 3.6 provides an overview of the methods followed by NGOs under different stages of the project.

It can be seen from the table that NGOs do not depend on one single method for any phase of the project. For instance, all the NGOs followed participatory methods for project formulation.But then, a few among them used survey method (29.6 %) along with participatory methods; 26 per cent of them

Table 3.6: Methods followed at Different Stages of the Project Cycle

Methods	*Formulation*		*Implementation*		*Evaluation*	
	No. of Res-ponses	*Percen-tage*	*No. of Res-ponses*	*Percen-tage*	*No. of Res-ponses*	*Percen-tage n=54*
PRA	54	100	41	75.9	51	94.4
Survey	16	29.6	—	—	14	25.9
LFA	14	25.9	11	20.4	1	1.9
Focus Group Discussion	8	14.8	37	68.5	—	—
Situation analysis study	3	5.6	—	—	—	—
Community meetings	—	—	27	50.0	—	—
Field visits	—	—	50	92.6	—	—
Group discussion	—	—	40	74.0	—	—
Camps	—	—	42	77.8	—	—
Campaign	—	—	36	66.7	—	—
Cultural programmes	—	—	32	59.3	—	—
Training	—	—	16	29.6	—	—
Seminar	—	—	15	27.8	—	—
Publication	—	—	12	22.2	—	—
Workshop	—	—	10	18.5	—	—
Counselling	—	—	9	16.7	—	—
External evaluation	—	—	—	—	11	20.4
Impact study	—	—	—	—	8	14.8
Case study	—	—	—	—	3	5.50

used Logistic Framework Analysis (LFA) along with PRA or survey method. Some of them (14.8 per cent) used Focussed Group Discussion. Thus, for project formulation, a considerable number of NGOs are found to use more than one method.

Similarly for implementing the projects too, NGOs adopted a variety of methods. Prominent among them are PRA (76.0%); field visit (92.6%); group discussion (74.1%); camps (77.5%); focus group discussion (68.5%); cultural programmes (59.3%) etc. For evaluating the projects, they largely depend on PRA (94.4%). The other methods used for evaluating the project are survey (25.9%); impact study (14.8%) and case study (5.5%).

Thus, the NGOs are using PRA as a major tool at different stages of the project. They are also using other methods to supplement the PRA method.

Problems faced

NGOs faced various problems at different phases of the project. Of the 54 NGOs, only eight reported that they did not face any problem at any stage of the project. Seventy eight per cent of the NGOs reported to have faced problems in formulating the projects. The main problems as reported by the NGOs are:

- paucity of secondary data (26.2%);
- lack of skill in drafting the project proposal (26.2%);
- problem in organizing and completing a survey (19.1%); and
- lack of support and cooperation from the village community (19.1%). (See Table 3.7).

Eighty five per cent of the NGOs reported that they faced problems in implementing the project. (See Table 3.8) Prominent among them are:

- lack of cooperation from the government (26.1%);
- natural factors like rain, drought, diseases etc., resulting in the death of cattle (23.9%);
- lack of support from local bodies (21.7%);
- factions and feuds in the village (17.4%);
- high degree of expectation from the people (17.4%).

Table 3.7: Problems Faced in Formulation of Projects

Problems	*No. of Responses*	*Percentage n = 42*
Lack of support and cooperation from the village community	8	19.1
Low women's participation	3	7.1
High degree of expectation from the people	3	7.1
Lack of secondary data	11	26.2
Unreliability of data on income, expenditure, savings	3	7.1
Illiteracy of villagers	3	7.1
Vested interests among the villagers	3	7.1
Problems in organizing & completing survey	8	19.1
Lack of skill in drafting the project proposal	11	26.2
Heterogeneous nature of villages	5	11.9
Improper identification of projects	3	7.1
Time consuming	3	7.1

Around one-third of the NGOs experienced problems in monitoring and evaluation of the projects. (See Table 3.9) These problems include:

— inability of the field workers;
— lack of proper feedback;
— lack of technical knowledge and guidance;
— time consuming nature;
— lack of seriousness among the beneficiaries; and
— problems in identifying events, case studies and specific methods.

Measurers taken to Overcome the Problems

Of the 54 NGOs, 46 reported that they have experienced some problems in one or two stages of the project. They also initiated some measures to overcome the problems. A measure that is

Table 3.8: Problems Faced in Implementing the Project

Problems	*No. of Responses*	*Percentage n = 46*
Lack of cooperation from Government	12	26.1
Natural factors – rain, drought, disease resulting in death of cattle	11	23.9
Lack of support from local bodies	10	21.7
Subsidy-oriented Government programmes have created a negative attitude on NGO projects	9	19.6
High degree of expectation from the people	8	17.4
Factions and feuds in the village	8	17.4
Lack of support and cooperation from the beneficiaries	6	13.0
Could not succeed in identifying IGP	5	10.9
Political interference	4	8.7
Time consuming	4	8.7
Want of adequate staff	3	6.5
Repetition of strategy, resulting in less participation	3	6.5
Interference from local bodies	2	4.4

adopted by a good number of NGOs is related to PRA. Around 61 per cent of the NGOs, who faced problems, reported that 'repeated use of PRA' helped them to overcome the problems. Similarly, 44 per cent of the respondents said that application of PRA from the beginning of the project has helped them to overcome many of the problems. (See Table 3.10) Other significant measures initiated by the NGOs include:

- combination of different methods (47.8%);
- orientation to beneficiaries (32.6%);
- frequent visits and meetings (30.4%);
- involving stakeholders at different stages of the project (30.4%);
- consulting experts and NGOs (26.1%);

Table 3. 9: Problems Faced in Monitoring & Evaluation of the Project

Problems	*No. of Responses*	*Percen-tage* n = 19
Inability of the field workers	5	26.3
Lack of proper feedback	3	15.8
Inadequate processing of information collected	3	15.8
Lack of technical knowledge & guidance	3	15.8
Time consuming (with a large number of beneficiaries)	3	15.8
Lack of seriousness among the beneficiaries	3	15.8
Problem in identifying events, case studies and specific methods	3	15.8

Table 3.10: Measures taken to Overcome the Problems in Project Formulation, Implementation, Monitoring and Evaluation

Measures taken	*No. of Responses*	*Percen-tage* n=46
Repeated use of PRA	28	60.9
Combination of different methods	22	47.8
Application of PRA from the beginning	20	43.5
Orientation to beneficiaries	15	32.6
Frequent visit to villages and meetings	14	30.4
Involving the stakeholder at different stages of the project	14	30.4
Consulting experts and NGOs	12	26.1
Periodic training to staff	11	23.9
Giving fair representation to different categories of people	11	23.9
Processing of data through computers	7	15.2
Discuss programmes with the government officials and leaders of local bodies	7	15.2
Improving accounting and record maintenance system	6	13.0
Streamlining the methods of data collection	5	10.9

— periodical training to staff (23.9%); and
— giving fair representation to different sections of the people in the community (23.9%).

USE OF PARTICIPATORY METHODS

Many development and research funding agencies have now imbibed PRA as an element of their institutional culture. All research and development proposals with rural development dimension now require explicit discussion of PRA to be acceptable to the agency concerned. PRA now seems to have some kind of 'flag of necessity' without which development oriented research and project vessels cannot acquire sea-worthiness certificate[3]. This might be one of the major reasons for the quick spread of PRA among NGOs, which predominantly depends on funds from government and other donor agencies. However, having taken up PRA as methodology, NGOs have made sincere attempts to incorporate the principles and philosophy of PRA in their projects, programme and activities. Further, they have been mainly instrumental in the spread and 'scaling up' of the PRA methods by honestly adhering to the very significant principle of PRA viz., 'sharing'.

It is against this background we have to look at the 'state-of-art-of PRA' among NGOs in Tamil Nadu. We now present an analysis of PRA methods as practised by NGOs, the benefits derived from such practice, problems encountered, and steps taken to overcome such problems.

Experience in PRA

PRA, as a methodology of development intervention work, was mostly evolved in late 80s. A few NGOs who were trained during this period in different settings have been responsible for pioneering and evolving new methods and techniques. They are also equally responsible for the spread of the PRA methodology. PRA as such is 10 years old in Tamil Nadu. The NGOs joined the school of PRA at different periods of time. Around 22 per cent of the NGOs have an experience of more than 8 years in the field of PRA. Around 46 per cent of them have had 6 to 8 years of experience;

Table 3.11: Years of Experience of NGOs in PRA

Experience in PRA	*No. of Respondents*	*Percentage*
< 2 Years	2	3.7
2-4 Years	5	9.3
4-6 Years	10	18.5
6-8 Years	25	46.3
8-10 Years	8	14.8
10 Years	4	7.4
Total	54	100.0

around 19 per cent have 4-6 years of experience. Thus, a vast majority of them have had long years of experience in PRA. (See Table 3.11)

Training in PRA

Mere experience in terms of number of years in the field does not help in the effective practice of PRA. Practitioners of PRA must have adequate training. Training plays a pivotal, prominent and critical role in practising PRA, as PRA is experiential and as it springs from action in the field, but not from academic analyses. Field-based participatory training is essential and it provides a sound base for the PRA practitioners to imbibe the culture of self-critical awareness, participation and sharing. On the other hand, a classroom based training in lecture mode totally neglecting affective domain is likely to undermine the very purpose of training and ultimately lead to low quality of PRA. Similarly "one-off training with no follow-up by trainers and by institutions is also a symptom and cause of low quality of PRA".[4] It is therefore important to see the number of trainings undergone by the NGOs and the duration of these trainings.

Barring 13 per cent of the NGOs, all the others were found to have undergone more than one training. In fact, little more than 25 per cent of the NGOs had undergone training for more than five times. (See Table 3.12)

Table 3.12: Number of Training Programmes undergone by Staff Members

No. of programmes	*No. of Respondents*	*Percentage*
One	7	12.96
Two	10	18.52
Three	16	29.63
Four	6	11.11
Five	7	12.96
More than 5	8	14.82
Total	54	100.00

Table 3.13: Duration of Training Programmes undergone by Staff Members

Duration of training progrmmes	*No. of Respondents*	*Percentage*
(No. of days) < 3 days	5	9.3
3-6 days	20	37.0
6-9 days	22	40.7
> 9 days	7	13.0
Total	54	100.0

Duration of Training

The duration of training is presented in Table 3.13. The duration here does not refer to total number of training days undergone by the trainee, but the duration in each training programme. The duration of training shows that many NGOs had undergone training for more than 3 days. Some had training (13.0%), which had a duration of more than 9 days. Most of the NGOs have, therefore, undergone training in PRA, among whom a majority had undergone more than one training.

Source of Training

The trainers and training institutions determine the quality of training and also subsequent practice of PRA in the field. Right from the beginning, NGOs played a very active role in imparting training in PRA. Significant among them are MYRADA (Bangalore), SPEECH (Madurai), ASA (Tiruchi), SEARCH (Bangalore), Women's Collective (Chennai) and Tamil Nadu Resource Team (Chennai). These institutions are known for their field-based participatory training in PRA. Majority of the NGOs reported that they have undergone training in one of these institutions. Among the NGOs, SPEECH tops the list followed by MYRADA, ASA, TNRT, and Women's Collective. Among the educational institutions, Gandhigram Rural Institute with 20 master trainers in PRA is known for imparting field-based participatory training in PRA. Around 45 NGOs reported that they have undergone training in GRI.

PRA Methods

PRA has a menu of methods. Practitioners have a choice of wide range of methods; many such methods are visual in nature. They may choose, pick and practise depending on the need. Further, they have the liberty to improvise the methods based on the situation or condition in a village.

All the organizations use more than one method in their work. About two per cent use three methods. A vast majority of them (76%) use 5 to 10 methods in their work while 11 per cent of the organizations venture into an extensive range of methods (more than 10 methods). (See Table 3.14)

Regarding the type of methods used, the NGOs are found to have used methods from a menu of 18 methods. When ranked from 'most' to 'least' used participatory methods, mapping and semi-structured interview scored very high; malady remedy analysis, SWOT analysis, causal diagram and access and control over resources at the bottom end of the range of methods practiced. The first one at the top is visual; and the second one is verbal which is used in combination with other methods like mapping, Venn diagram, wealth ranking, seasonal calendar etc. Other methods with high scores are group discussion (70.4%), seasonal calendar (68.5%), ranking exercises (59.6%), and venn diagram (48.1%). The least used methods include linkages, focus group

Table 3.14: Number of PRA methods used by NGOs

No. of methods used	*No. of Responses*	*Percentage*
One	—	—
Two	—	—
Three	1	1.9
Four	2	3.7
Five	3	5.6
Six	12	22.2
Seven	11	20.4
Eight	10	18.5
Nine	5	9.3
Ten	4	7.4
Eleven	3	5.6
Twelve	1	1.9
Thirteen	1	1.9
Fourteen	1	1.9
Total	54	100.0

discussion, decision making matrix, case studies and shared presentation. (See Table 3.15).

The methods used by the NGOs by and large, seem to indicate that they (i) use a combination of methods and (ii) go in for sequencing. Combination of methods would mean combining two or three methods at a time to get accurate, cross-checked data. For instance, while undertaking a transect, the participants normally combine methods like transect, direct observation, semi-structured interview techniques and shared presentation. Such combination help the PRA practitioners get a variety of rich, relevant and accurate data.

Sequencing means practice of one method naturally takes the practitioner to another method in order to have an in-depth analysis of a problem or situation. Sequencing of participatory methods is

Table 3.15: Use of Participatory Methods by NGOs

Type of Methods	*No. of Responses*	*Percentage*
Mapping	54	100
Semi-structured interview	54	100
Group Discussion	38	70.4
Seasonal calendar	37	68.5
Ranking exercises	32	59.4
Venn diagram	31	57.4
Timeline	27	50.0
Trend change	26	48.1
Transect	26	48.2
Linkages	14	25.9
Focus Group Discussion	13	24.1
Decision-making matrix	12	22.2
Case studies	10	18.5
Shared presentations	10	18.5
Access and control over resources	9	16.7
Causal diagram	7	13.0
Malady remedy analysis	5	9.3
SWOT analysis	6	11.1

both powerful and purposeful.[5] Robert Chambers[6] has very vividly described the power of sequencing. They are as follows :

— the commitment of participants increases, making further action more likely, more spontaneous and more sustainable;

— sequences triangulate, and reveal errors or omissions in earlier presentation;

— different activities interact cumulatively, each activity adding a dimension and details which qualify and enrich others, so that taken together the whole becomes more than the sum of its parts; and

— all concerned learn through the process, through local people sharing what they know, through observation and through analysis.

For example, social mapping shows the demographic and social characteristics, leading to an identification of key informants, and then to discussion with them and analysis by them. Similarly social mapping may lead to wealth or well-being ranking and subsequently to livelihood analysis; a participatory resource map leads to planning transect walks in which villagers who made the map act as guides for outsiders. The transects in turn lead to the identification and discussion of problems and opportunities, which then lead to listing and ranking options on 'best bets'.

All the NGOs cannot be quoted as practising the concept of 'sequencing'. Only the sufficiently trained and consistent practitioners can go in for sequencing. Sequencing also depends on the type, nature and quality of data required.

Practical Applications

Applications of PRA approaches and methods have proliferated and continue to multiply. PRA methods are applied in different fields. The NGOs under report are found to have applied the methods in different areas. When ranked from 'most' to 'least' applied areas of participatory methods such as women development (96.3%), financing/micro credit (87.0%), rural water supply and sanitation (83.3%), animal husbandry (83.3%), micro enterprises (74.1%), and health (74.1%) have scored high with majority of NGOs applying PRA in the above fields. Very few NGOs applied the concept of PRA in policy initiative and change (9.3%), and tank rehabilitation (11.1). In between high and low categories lie areas like agriculture (46.3%), forestry (40.3%), education (42.6%), training (27.8), human rights (27.8%), organization building (18.5%) and watershed management (18.5%). It is interesting to note that NGOs have started applying the PRA concept in lesser known but more important areas like organization building, policy initiative and change and human rights. (See Table 3.16)

Table 3.16: Application of PRA by NGOs

Areas	*No. of Responses*	*Percentage*
Women development	52	96.3
Financing / Micro-credit	47	87.0
Rural water supply and sanitation	45	83.3
Animal husbandry	45	83.3
Micro-enterprises	40	74.1
Health	40	74.1
Agriculture	25	46.3
Forestry	25	46.3
Education	23	42.6
Housing	20	37.0
Training	15	27.8
Human rights	15	27.8
Horticulture	10	18.5
Watershed management	10	18.5
Organization building	10	18.5
Tank rehabilitation	6	11.1
Policy initiative and change	5	9.4

Benefits of using PRA

The benefits of PRA, as experienced and experimented, are many. All the stakeholders in PRA benefit. The benefits/advantages in general are listed below :

- — participation of people is quickly secured and ensured
- — less costly (expensive)
- — less time-consuming
- — flexible and adaptable
- — local support and help for organizing appraisal
- — all categories of people articulate their problems, concerns

and priorities
- participation of women in large number
- visual presentations of fact and data about the people make them realize their conditions very easily
- sequencing of methods
- building up rapport between the outsiders and insiders
- building up confidence in the minds of participants
- empowering the people.

The benefits listed above are only indicative and not exhaustive. These are mainly drawn from the experience of the practitioners. All these are general in nature. But then, what do the NGOs, as consistent practitioners of PRA for a sufficiently longer period of time, perceive as the benefits of PRA? The responses are presented in Table 3.17. The NGOs have reported as many as 19 benefits. The most prominent among them, as perceived by more than 50 per cent of the respondents, are :

- clear assessment of the existing situation at micro level (94.4%);
- collection of wide range of information (94.4%);
- creation of closeness between the outsiders and insiders (92.6%);
- less time-consuming (90.7%);
- cost effective (88.9%);
- accuracy of data (87.0%);
- anybody irrespective of age, sex or literacy status can use PRA (87.0%);
- collection of qualitative data (85.2%);
- helps in accepting other ideas leading to mutual respect(66.7%); and
- helps to identify focus groups (61.1%).

The benefits, as expressed by the NGOs, coincide with the advantages as harvested from their experience.

Table 3.17: Benefits derived from using PRA for the Outsiders

Benefits	*No. of Responses*	*Percentage*
Wide range of information could be collected	51	94.4
Clear assessment of the present situation at micro level by insiders and outsiders	51	94.4
Builds closeness between insiders and outsiders	50	92.6
Less time-consuming	49	90.7
Cost effective	48	88.9
Collection of accurate data	47	87.0
Anybody irrespective of age, sex, literacy status can use PRA	47	87.0
Collection of qualitative data	46	85.2
Helps in accepting others ideas and mutual respect develops	36	66.7
Helps to identify focus groups	33	61.1
Continuous practice of PRA brings attitudinal & behavioural change	32	59.3
Collection of relevant data only	31	57.4
Large number of beneficiaries could be covered with less effort	30	55.6
Various rural issues, problems and social life could be studied better	27	50.0
Responsibility shared	27	50.0
Has brought about a near-elimination of the feelings of hierarchy among the staff	26	48.2
Arriving at decisions acceptable to all	20	37.0
The process is intensive in terms of discussion and participation	18	33.3
Leads to action	17	31.5

Benefits to People

The benefits of PRA to the people as perceived and reported by the NGOs are presented in Table 3.18. The most important and significant benefits are :

- — creates self-awareness about the conditions and situations in which they live (96.3%);
- — helps participants for self assessment (94.4%);
- — participants freely articulate their needs/problems and aspirations (90.7%);
- — PRA methods are easy to understand and simple to practice (87%);
- — assessment of the existing situation enables the participants to visualize what should be done to improve the situation (85.2%);
- — even passive members participate in the programme in course of time (77.8%);
- — effective implementation of the project with active participation of people (75.9%);
- — results in skill development in analysis and planning (74.1%);
- — participants can have a document on themselves (74%); and
- — facilitates participatory decision making process (68.5%).

The benefits as derived by the outsiders and by the insiders, in tandem, are likely to have a perpetual impact on the people and also on the outsiders. It is this multitude of benefits, which made the PRA the most popular and effective method of appraisal, analysis, action and evaluation.

Problems in practicing PRA

Practice of PRA is not free from problems. The practitioners in the field do face lot of problems especially in the initial stages. The problems as encountered in practising PRA are classified into four categories. They are: (i) people-based; (ii)team-based; (iii) approach based; and (iv) tool based. The NGOs have been asked to state

Table 3.18: Benefits derived from using PRA for the People

Benefits	*No. of Responses*	*Percentage n = 54*
Creates self-awareness about the situation and condition in which they live	52	96.3
Help the participants for self-assessment	51	94.4
Participants freely express their needs/ problems and aspirations	49	90.7
Easy to understand and simple to practice	47	87.0
Assessment of the existing situation enables the participants to visualize what should be done to improve the situation	46	85.2
Even passive members participate in the programmes in course of time	42	77.8
Effective implementation of the project with active participation of the people	41	75.9
Participants have a document of themselves	40	74.1
Leads to skill development in analysis and planning	40	74.1
Facilitates participatory decision-making process	37	68.5
Capacity building among the weaker sections through training, discussion and periodic meeting	29	53.7
New type of leadership is created among the marginalised section	27	50.0
Instils self confidence among the people	26	48.2
Villagers serve as resource persons among the educated	23	42.6
Enhance the planning skills of people	23	42.6
Participation gets internalized	21	38.9
Use of participatory methods in training ensures the participation of all	21	38.9
People start recognizing the social and environmental dimensions of the project	12	22.2

the problems faced by them under each category. The responses of NGOs are consolidated and presented here one by one.

People based problems: PRA is basically people-centered exercise. It is pro-poor. It is totally free from gender-bias. Its primary focus is to involve different sections of the community in a village and enable the people to make an objective analysis of their conditions. The ultimate aim, as repeatedly claimed by the practitioners, is to make the people self-supporting, self-reliant and finally to empower them. This presupposes participation of the people. It must not be a mere physical presence; but active participation at different stages of the appraisal. Does it happen? Do people actively participate? What are the problems NGOs face in enlisting people's participation? The responses of the NGOs are presented in Table 3.19.

A major problem as reported by 98 per cent of the respondents is "high expectations from the people" during the Participatory Rural Appraisal. This invariably happens especially in the case of "organized PRA" where the participants continuously work with outsiders in analyzing their conditions on specific problems. Such a kind of analysis creates a feeling among the participants that it would bring some welfare oriented action at a later stage. Launching an organized PRA exercise involving arrangements for the stay of a group of outsiders in a village represents a visible commitment on the part of a project to a particular community. This raises the expectations of people. This is true, as the practitioners of PRA have often reported 'dependency syndrome' as the most serious problem in PRA exercise.

The second important problem as reported by 89 per cent of the respondents is that social/geographic/distance within a village elicits non-cooperation from the people. Mosse says that PRAs are not likely to be equally accessible and open to all sections of the community.[7] This is especially so when the village is big with different categories of people living in different parts of the village. This is supported by the fact that around two-thirds of the respondents reported that in a heterogeneous community, a section of the people do not participate actively.[8]

Table 3.19: People based Problems

Problems	*No. of Responses*	*Percentage n = 54*
People tend to develop high expectations	53	98.2
Social/geographic difference/distance within a village elicits non-cooperation from one section	48	88.9
Domination and vested interests of a few affects the quality and reliability of information	42	77.8
Women in some communities do not participate at all	39	72.2
In a heterogeneous community, a section of the people do not participate actively	36	66.7
Youth are initially hesitant	18	33.3
People have a tendency to conceal certain information	16	29.6
People tend to expect immediate results	12	22.2
Psychological unity of the community some times camouflages the information	9	16.7

Yet another people-based problem is domination and vested interests of a few which affect the quality and validity of the information (77.8% responses). This is true to a greater extent because "the dominant views will tend to dominate"[9]. Further, consensus arrived at in the PRA setting may be forced on the villages by the village leaders.[10] Murphy, for instance, talks about established local political procedures for consensus formation. He describes how village elders create space for people to express difference of opinion in public meetings, and how through managing this space the decision made generally favours village elders.[11]

Yet another significant problem is that women in such communities do not participate at all (72.2% of the responses). This is mainly because of social constraints. PRA usually takes place in public places and in the presence of outsiders. Women in such situation may not participate. Further, women from high caste groups do not participate whereas women from low caste groups participate not by physical presence alone but actively participating in the deliberations. The other problems as reported by the respondents are: initial hesitation on the part of the youth; tendency to conceal information (29.1%) and psychological unity among the community tend to camouflage the information (16.7%).

Team based Problems: There are two types of teams in PRA exercises. They are: (i) team from outside the village (known as outsiders/researchers/facilitators); this team should be multi-disciplinary in nature representing different disciplines; (ii) the team from within the village; it must be a group representing different sections of the community including women. These two teams should work in close collaboration with each other for better results.

Failure to constitute a good representative group of people from the village was reported by 91 per cent of the NGOs as the main team based problem. This problem of not reacting all interest groups persists. The group from the village may not have fair representation of the women, the landless, ethnic minority and the poorest. The literates, the rich, the landlords and those with vested interest take over the talking, the drawing, the opinion domination and the poor are leftout.[12]

Yet another team based problem as reported by around 80 per cent of the NGOs is 'trying to get to the people through a contact person in the village'. This has serious implications especially when the contact person is not neutral. This may most probably lead to biases and subjectivity in conducting PRA type of enquiry. Another problem is inability to review and evaluate the PRA exercise (79.6%). Review and evaluation is a sort of self-critical examination, which is a very cardinal philosophy of PRA. This ultimately leads to tuning and fine-tuning of the process of PRA for doing PRA in a better way. Hence, this problem has to be viewed very seriously.

Little more than two-thirds of the respondents felt that there is a tendency on the part of the team to go in for familiar PRA exercises. This is not a very serious problem; but it restricts and inhibits innovations in PRA. Absence/less number of women in

Table 3.20: Team based Problems

Problems	*No. of Responses*	*Percentage n = 54*
Failing to constitute a good representative group of people from the village	49	90.7
Trying to get to the people through a contact person in the village	45	83.3
Inability to review and evaluate the PRA exercises	43	79.6
Tendency on the part of the team to go in for familiar PRA exercises	37	68.5
Absence/Less number of women members in the team	19	35.2
Failing to establish adequate rapport with the people	11	20.4
Lack of coordination and inadequate preparation among the PRA team	9	16.7

the team[13] (35.2%) and failure to establish adequate rapport with the people (20.4%) are other team-based problems. (See Table 3.20).

Approach based problems: PRA is basically an approach for learning about rural life and conditions. The 'approach' is related to attitude and behaviour of the outsiders. The methods (PRA) are likely to bring desired results only when the attitude and behaviour of the team is positive. The attitude and behaviour part of the PRA emphasizes qualities like openness, flexibility, transparency, non-domination, listening to people, avoiding lecture and so on. The major approach based problem is failing to conduct plenary session in the village (90.2%). Plenary is very important especially when more than one group in a village analyses the condition of village using different exercises. The outcome of such exercise done by different groups in the village need to be presented and shared among the participants for better understanding and for obtaining a wholistic picture about the village. 'Failure to conduct plenary session' results in sacrificing 'sharing' a very significant principle of PRA.

Around 78 per cent of respondents reported that "attempts were not made to elicit the views of passive participants". PRA, in fact, is mainly evolved to see the biases in rural development tourism are offset. The core philosophy of PRA is to empower the weak, and the powerless. The weak and the powerless are normally passive. They should be empowered by enabling them to articulate and participate. Yet another problem as reported by the NGOs is that "trying to adopt the same approach in all the villages" (46.3%). Villages and people vastly differ from each other. Hence following the same approach in all the villages is likely to hinder the process of doing PRA.

Wrong choice of the venue of appraisal (27.8% response) will restrict the participants. Temple or chavadi (a public meeting place in a village) normally restricts the participation of women in PRA. Very lengthy sessions (25.9% response) without minding the inconvenience and hardship of the participants especially the women are likely to bring down the level of participation. The other approach-based problems as reported by the NGOs include "not staying overnight in the village" (20.4%) "lack of transparency on the part of outsiders" (13.0%), "lack of flexibility" (11.1%) and "imposing blueprint upon the villagers" (9.2%). (See Table 3.21).

Tool based problems: Tools or methods is one of the three legs of PRA—the other two being attitude and behaviour and sharing. A very significant advantage of PRA tools is that they are mostly visual exercises, which facilitate high degree of participation of the people in PRA exercises. The visual presentation and analysis have several other advantages too. With visual analysis the relationship changes; the outsiders are conveners and facilitators and the insiders serve as actors and catalysts. The outsiders hand over control and insiders determine the agenda, categories and details. The media and materials are often those of insiders. Information is built up cumulatively and crosschecking is automatic. Often several or many people are involved and therefore knowledge overlaps. Discussion is lively because everyone can see what is being said. It can then be

Table 3.21: Approach based Problems

Problems	*No. of Responses*	*Percentage n = 54*
Failing to conduct plenary session in the village	49	90.7
Not trying to elicit the views of passive participants	42	77.8
Trying to adopt the same approach in all the villages	25	46.3
Wrong choice of the venue of appraisal	15	27.8
Very lengthy sessions	14	25.9
Rushing to complete the exercise early	12	22.2
Not staying overnight in the village	11	20.4
Identification and collection of focus groups	9	16.7
Lack of transparency	7	13.0
Lack of flexibility	6	11.1
Imposing a blueprint upon the villagers to complete the exercise early	5	9.3

diagram rather than the people who are interviewed. It can empower the weak and the disadvantaged. Visual literacy is independent of alphabetical literacy and appears to be near universal. Visual diagramming is an equalizer when it is done using accessible and familiar medium of ground. All who participate—the children, women, men, poor, rich, illiterate, literate—can similarly understand what is being shown.[14]

Tools and methods could be improvised and used depending on local situation and condition. There is flexibility. The improvising of methods, flexibility and adaptability have mostly contributed to the rapid spread of PRA.

Some of the problems stated by the NGOs are related to visual exercises. For instance, 85 per cent of the respondents reported that many exercises, unless done sector-wise, became very difficult. (e.g. mapping) This is all the more so especially in a big

Table 3.22: Tool based Problems

Problems	*No. of Responses*	*Percentage n = 54*
Many exercises unless done sector-wise, becomes very difficult	46	85.2
Occupational diversity /heterogeneity in the village causes certain exercises becoming difficult and less meaningful	42	71.8
Limitations in bringing out visual presentations in some exercises	40	74.1
Large size of the village(s)	35	64.8
Problems in gathering and using locally available materials	22	40.7
Trying to bring in many variables dilute the exercise	19	35.2
Subjectivity and vested interests creep in among the participants in a PRA exercises	7	13.0

village, which is heterogeneous in nature. When the number of households exceeds 250, it is better to go in for street-wise or sector-wise mapping of the village. The NGOs under study (74.1% response) have had limitations in bringing out visual presentation in some exercises like trend analysis, time line, livelihood analysis and so on. Such limitations may force the participants to resort to alphabetical literacy, which is likely to inhibit and restrict the participation of people.

Yet another tool-based problem reported by the NGOs (77.8%) was that in the event of occupational diversity and/or heterogeneity in the village, certain exercises become difficult and less meaningful -the examples being wealth ranking and seasonality. This is because the criteria for wealth/well being ranking differ across different occupational groups and community. Similarly the seasonal calendar of people regarding income, expenditure, savings, debt, diet, food consumption, fodder, fuel, and migration may exhibit differences and disparities among different sections of the community. Other problems as stated by the NGOs include:

- problem in gathering and using locally available material (40.7%);
- trying to bring in many variables which dilute the exercise (35.2%); and
- large size of the village (64.8%).

Solution to Problems

The NGOs faced certain problems while organizing PRA exercises especially at the initial stages. The problems are not insurmountable. The NGOs have largely overcome them through repeated practice of PRA in different settings. The solutions to overcome the problems are not theoretical but practical. The solutions to different problems as offered/followed by the NGOs are summarized and presented. Each solution offered by the responding NGO is very important and hence we have not gone in for prioritizing the solutions in terms of the number of responses. The solutions highlighted below are those mentioned most often. They are presented under different headings.

Solutions to People based Problems

The steps taken by the NGOs to overcome the problems arising out of lack of proper representation are: (i) changing the location; (ii) improved rapport building; (iii) repeated visits; and (iv) repeating the exercises itself in a different setting.

For enhancing the participation of women in PRA, the measures taken are: (i) choosing locations where women folk are quite used to assemble; (ii) including women members in the team of facilitators; and (iii) requesting the menfolk to allow women to participate in the PRA.

Regarding high expectations from the participants, the NGOs are of the view that so long as the NGOs and its personnel do not nourish the dependency syndrome, there is nothing wrong in the people coming up with expectations as a result of PRA. Due care need to be taken to channel their expectations into the participatory techniques of seeking solutions. NGOs should maintain transparency in their dealings.

The expectations have to be carefully managed. If PRA is seen as a condition for grants, the exercise will become a farce. We have to spend time with the villagers before the exercise to explain its purpose: 'a learning experience for all'; 'a participatory exercises for all'; 'the beginning of their organized efforts towards sustainable development based on their strengths, traditional values and management systems'. But commitments are often required; these, however, are not in terms of funds, but of instilling confidence in the villagers that outsiders will continue to be with them and to support them, to build on their strengths and to overcome their weakness and to develop an appropriate strategy to achieve their goals. The commitment therefore is more in the form of 'personal involvement' with their effort than of monetary inputs.

Solutions to Team based Problems

The measures taken by the NGOs to overcome the team-based problems are listed below:

- A few preliminary visits to the village are not just enough to establish rapport and gain confidence of the people. PRA is not an exercise to establish a 'point of entry'. It needs preparation, primarily through personal contact with the village over a period of time. Hence, the NGOs are reported to have firmly established contact and then went in for PRA.
- Only those villages where PRA exercises are followed by action projects are selected for doing PRA exercises.
- It is better that the PRA team from the change agent arranges for the program in the village concerned directly without using a contact person.
- We should ensure that there are women members in the team of PRA facilitators in order to make the team gender balanced.[13]
- The PRA team should not go to any village with a pre-conceived idea, but should design a programme only based

on the demands of village situations.

- Precaution should be taken in the preliminary meeting that no member should intervene in the middle of the programme and deviate the focus of PRA team already at work.

Solutions to Approach based Problems

The PRA Team must avoid going into a village without any idea about the situation prevailing in the village. Even secondary data could be used to get a clearer picture of the village. The PRA team should work out a fitting approach in line with the assessment of the village.

- The team should learn to be relaxed and avoid rushing. However, if the villagers seem to rush for whatever reasons, the team should withdraw slowly but before winding up, the next sitting is to be finalized with the group.
- Plenary sessions must be considered as imperative before following the series of PRA exercises. However, if plenary sessions could not be conducted immediately it may be organized later. The plenary should follow without inordinate delay.
- Villagers should not be coerced into a compromise regarding the time schedule of PRA exercise.
- The grassroots level personnel of an agency should consider the advantages of night-halt very seriously, in case they do not happen to be staying in the village.
- However urgent is the topical relevance of an issue of an agency in its own rural work, a prioritized common issue emerging through PRA must be attended to at least to some extent.
- Despite the importance to be attached to key informants, the facilitator must see to it that it does not lead to their domination.
- Imposing of blue prints on the villagers must be avoided at any rate.
- The choice of venue in PRA exercises must be left to the

villagers even if they happen to choose a wrong one. It is better that the people are let to realize their error, than the facilitator evaluating the same and suggesting an alternative.

- Facilitator should not be too ready to offer clues in a PRA exercise because this may kill the initiative of the villagers.
- PRA team may have to keep making efforts rather untiringly for bringing together people for a 'general discussion' in view of their importance in certain themes.
- PRA philosophy obliges the team members to spare no efforts in convincing the people on the long-term impact of PRA approach.
- The meeting should always be conducted in an 'environment of safety' which is possible only where it is guided by 'dialogue' not by 'debate'. In debates, people have to defend themselves and the environment is often threatening. In dialogues, the environment is safe and participants are invited to express their understanding and listen to each other to build together 'richer pictures'.[14]

Solutions to Tools based Problems

Efforts must be made to bring in 'ground maps' and visual representation as far as possible. However, use of visual need not be pushed to the extent of making the participants uncomfortable if they have chosen otherwise.

The problem related to using locally available materials could be avoided through limiting the number of variables in an exercise. This may necessitate using other type of mapping to accommodate more variables in the analysis.

In a heterogeneous situation/village, PRA exercise may have to be repeated on several occasions and several locations for accurate, wholistic data.

Conclusion

The focus of this chapter is to present the profile of NGOs and to analyze PRA methods as practised by NGOs, problems encountered and steps taken to overcome such problems. The analysis is based on the responses received from 54 practitioners through mailed questionnaire.

The NGOs interviewed through questionnaire have long years of experience in voluntary work. Many of them have wider area of operation covering more than 50 villages. The objectives of NGOs mainly centered on empowering women and marginalised sections of the community through awareness generation, organizing, financing and initiating income generation programmes. The activities undertaken by them are well in tune with the objectives laid down. The beneficiaries are mostly economically downtrodden and socially marginalised. The methods followed in their project largely reflect their participatory nature. Many NGOs are reported to have faced problems at different stages of the project, which are mostly internal in nature. Such problems, the NGOs reported, have been tackled mainly by using participatory methods.

A vast majority of the NGOs have long years of experience in PRA. A few of them are 'first generation practitioners of PRA' exposed to field based participatory training. NGOs like MYRADA, SPEECH, ASA, SEARCH, WOMEN'S COLLECTIVE, TNRT and educational institutions like GRI are the major centres of training for PRA to the reporting NGOs. Regarding the use of PRA methods, all the NGOs use more than two PRA methods in their work. Three-fourths of them use 5 to 10 methods. The methods, which are used by majority of NGOs, include mapping, SSI, seasonal calendar, ranking exercises and Venn diagram. The methods that are least used (in terms of number of NGOs reporting) are malady-remedy analysis, SWOT analysis, causal diagram, and access and control over sources. They use a combination of methods and sequencing for better results. The major areas of application are women and development, micro-credit, rural water supply and sanitation, animal husbandry, micro-enterprises, and health. PRA is also applied in organization building and policy initiation and change.

The major benefits the researchers/outsiders derived from the application are: need assessment of the existing situation at micro-level, collection of broad spectrum of information, rapport between insiders and outsiders, and cost effectiveness in terms of time and money and accuracy. PRA is beneficial for the villagers too. The benefits are: self awareness, self-assessment, free and frank articulations of their needs and problems, future visioning, participation of all, and skill development in analysis and policy.

The people based problems as reported by the NGOs are: dependency syndrome, non-cooperation from the people due to social/geographic distance, domination of vested interests affecting the quality and validity of information, and less/non participation of women. Major team based problems are failure to constitute a representative group and ability to review and evaluate the PRA exercise. The approach-based problems are failure to conduct plenary session, failure to make the passive participants to speak, and adopting same approach in all the villages. The tool-based problems include big and heterogeneous nature of the village make practice of certain exercises difficult.

Solutions were also offered for the problems. The main solutions are related to continued and consistent use of participatory methods and strict adherence to the process and philosophy both on the field and off the field.

NOTES AND REFERENCES

1. Robert Chambers, *Whose Reality Counts: Putting the First Last,* Intermediate Technology Publication, London, 1997, p. 114.
2. PRA Unit, *PRA at GRI: Looking Back at the Trodden Path,* Gandhigram Rural Institute, Gandhigram, 1996.
3. Paul Richards, "Participatory Rural Appraisal: A quick and dirty Critique", *PLA Notes 24,* October 1995, p. 13.
4. Absalom, Elkanah, *et al.,* "Sharing our concerns and looking to the future", *PLA Notes 22,* February 1995, p. 9.
5. Robert Chambers, *op.cit*, p. 136.
6. Robert Chambers, "Participatory Rural Appraisal: Analysis of Experience", *World Development*, Vol. 22, No. 9, 1994, p. 1257.
7. David Mosse, "Authority, Gender and Knowledge: Theoretical Reflection on Participatory Rural Appraisal", *Economic and Political Weekly,* Vol. XX, No. 11, March 18, 1995, p. 573.
8. In one of the PRA workshops organized by GRI-PRA team in Narasingapuram village near Gandhigram, a particular community dominated the entire proceedings of the workshop for all the three days. The other section of the community did not actively participate. This happened in

spite of the earlier request to the leaders to give fair representation to all the communities of the village in the PRA proceedings. See PRA Unit, *Rural Planning through People*, GRI-PRA Series, Gandhigram Rural Institute, Gandhigram, May 1992, p. 14. (Unpublished)

9. David Mosse, *op.cit*, p. 573.
10. A few village leaders are likely to dominate the discussion in spite of strict adherence of ground rules like 'avoid concentrating on lead informant'. For instance, while facilitating a seasonal calendar in Oothupatti villages near Gandhigram, two leaders of the village were providing information though there were 12 participants in the village. As true practitioners of PRA, we cautiously shifted our focus from the lead informants to other passive informants and asked them to share their views and opinion. The lead participants did not like this and said, "We speak on behalf of the group. Our voice is the voice of the group". We replied that "everybody should be heard". They did not accept this and left in the middle of the exercise.
11. Murphy William, "Creating the Appearance of Consensus in Mende Political Discourse" *American Anthropologist*, 1, 1992, pp. 22-41.
12. Britha Mikkelsen, *Methods for Development Work and Research*, Sage Publications, New Delhi, 1995, p. 198.
13. Absence of women member in the outside team is a factor that restricts the participation of women from the village. A team of PRA practitioners went for a preliminary visit to a Vedapatti village near Dindigul (TN) and discussed with the people about the ensuing PRA workshop and sought the participation of people especially the women. The women as well as men insisted that unless the outside team (GRI-PRA) has women representative, the women participants from the village would not participate.
14. Robert Chambers, "Participatory Rural Appraisal: Analysis and Experience", *World Development*, Vol. 22, No. 9, 1999, pp. 1263.

4

USE OF PARTICIPATORY METHODS IN PROJECT CYCLE

The Experiences of NGOs

Participation is rapidly becoming a catch-all concept, even a cliche.[1] It has been increasingly emphasized in all rural development projects. The Government too, after going through seven Five Year Plans, finally recognized the significance of participation and stated in the Eight Five Year Plan thus: "People's initiatives and participation should be made a key element in the process of development instead of people being passive observers"[2] Participation is of different types. The level and degree of participation of the people varies. It depends largely on how, where, at what level, the researcher/development agencies intend to involve the people in projects. Biggs has sub-divided participation into four types:

- *Contractual* where people contracted into the project of researchers/development agencies take part in their enquiries or experiments;
- *Consultative* where people are asked for their opinions and consulted by researchers/development agencies before interventions are made;
- *Collaborative* where researchers and local people work

together on project designed, initiated and managed by researchers;

- *Collegiate* where researchers/ development agencies and local people work together as colleagues with different skills to offer, in a process of mutual learning where people have control over the process.[3]

The last one is the highest form of participation and would mean the active involvement of people at every stage/phase of the project. The emphasis here is very much on bottom-up approach with a clear "focus on locally defined priorities and local perspectives."[4] PRA preaches and practices 'collegiate' type of participation. Participation at this stage seeks "knowledge for action"[5]

Many development agencies (mostly NGOs) including the donor agencies started grounding the concept of 'collegiate' type of participation in their development project—the core elements being process orientation, bottom-up approach, and participatory methods. The advantages of such approach are many. They may be briefly stated:

- data are collected, analyzed and tested directly by users;
- problems and their solutions are tried out at the planning process itself;
- appreciation of the stakeholders' problems, cultural context, possibilities and conditions of change is enhanced;
- weaknesses and strengths are perceived by the participants directly; and
- People's motivation to participate directly in the project implementation is enhanced through their participation in decision-making and better understanding of the problems.

NGOs working at the grassroot level quickly recognized the 'power' of participatory methods in micro planning. The advent of PRA made a perceptible impact on the development strategy adopted by the NGOs who were mostly depending on top-down approach with survey as the major method of information gathering for planning. The scenario has changed a lot and PRA has virtually taken over the place of survey in many of the NGOs as PRA has a built-in mechanism to encourage the participation of local people in the process of development by: investigating local insights gained by interfacing with common men, resulting in more effective research information being located; and involving local people in research and design of socio-economic endeavour so as to lead to increasing commitment and empowerment of common people.[6] The experiences of hundreds of thousands of practitioners and the literature on participation have outlined the gains of PRA. They are: (i) local knowledge is mobilized, (ii) local resources and energy are used (iii) participation is enhanced (iv) actions are better adapted and more sustainable. (v) adaption is fast and fitting to changes in local conditions (vi) costs are reduced (vii) central administration is reduced (viii) political problems and conflicts are resolved locally.[7]

Having comprehended the relevance, significance and gain of PRA in bottom-up micro planning, we now present the cases of four NGOs who have applied the concept of PRA at different levels of the project. In the third chapter, we provided an overview of PRA as practiced by NGOs. Here, we attempt an in-depth analysis of the experience of four NGOs. We also make an attempt to bring in the views/opinion of the people who are practicing PRA. We feel this as quite essential, as the available literature does not contain any feedback from different community members of what PRA is all about or how it compares with other existing approaches to problem analysis.[8]

The chief questions that we may ask in analyzing the place of PRA in NGOs are:

- Why do they apply PRA in their projects?
- How do they view PRA as a methodology?
  - — for gathering information
  - — for formulating projects
  - — for planning their activity
  - — for executing and evaluating their programme
- What is the relevance and efficacy of participatory methods at different stages of the project?
- Have they felt the gain of PRA as experienced by their fellow practitioners?

- What do people-the partners in development-view about the whole lot of participatory methods?
- Have participatory methods empowered the poor? If so, how and to what extent?
- Have they attempted any innovations? If so what are they? Do they find them useful and relevant to their development work?

We now present the cases. In presenting the cases, we give a brief account of the practice of PRA by the selected NGOs followed by presentation of cases based on field experience and finally draw inferences from cases. The cases are based on interaction with NGO heads, project coordinators, field workers and the villagers who participated in the PRA exercises. The inferences drawn from the case studies fully reflect the views of the different stakeholders in the participatory development process, as they are the inferences of the people and not the researchers.

OUTREACH, BANGALORE

Background

OUTREACH is a non-government development organization-established in May 1993 in response to the needs of the communities living in drought prone areas of South India and also to the growing needs of training in the NGO sector and in Government. The idea was to focus on small NGOs working in drought prone regions in South India, particularly those engaged in natural resource management activities like watershed development, wasteland management, forestry, agro forestry, tank rehabilitation and so on. The organization has covered 870 villages in three States viz., Andhra Predesh, Tamil Nadu, and Karnataka.

OUTREACH has been practicing PRA in all its development work and initiatives right from its inception. It has applied the concept of PRA in all its activities which include watershed management, forestry, horticulture, animal husbandry, fisheries, tank rehabilitation, rural water supply, environment, health, sanitation, micro-credit, micro-enterprises, energy, poverty eradication, gender education, training, organization building, policy initiative and change. The organization applies PRA throughout the project cycle, which include situation appraisal, project formulation, planning of schemes and activities, budgeting, implementation, monitoring and evaluation.

OUTREACH has also used participatory methods in building organizations. PRA has largely helped in forming the *Sangas*. *Sangas* were never thrusted on the people. They all evolved through discussion and dialogue. Sangas in turn felt the necessity of having their own federation called Cluster Level Association (CLA). CLAs, in turn, have found their federation known as Block Level Association (BLA). (See Exhibit 1)

All the organizations from village level (Sanga) to Block level (BLF) are run by the members. The members run CLA. CLA serves as resource person and provide training to the members of Sangas in the area of maintenance of accounts, auditing of accounts, project preparation and monitoring. The trainers are both at Sanga

EXHIBIT 1: PRA in Institution Building

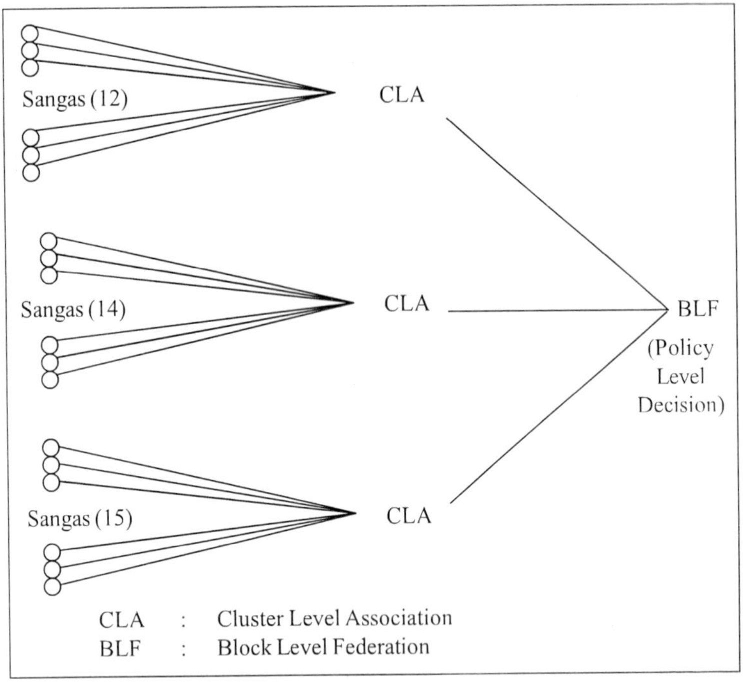

and at CLA level. Trainers are identified in the area of savings and credit, community organization, PRA, accounts and audit. The women so identified normally impart training to fellow members.

CASE I

Continuum between Appraisal and Planning: The Case of Usthanapalli

Appraisal is not just for the sake of knowledge. Any appraisal based on participatory methods, the practitioners of PRA insist, must lead to action. The staff of OUTREACH said that they don't stop with appraisal. It automatically takes them to the next stage viz., planning. We present a case where appraisal of situation in a village resulted in planning certain activities.

OUTREACH drew many methods from PRA menu; but the methods used quite often are (i) chappathi diagram in different forms; (ii) matrix scoring; and (iii) 100 seeds scoring. The reasons are: (i) villagers are quite at ease in handling the exercise and could easily practice; (ii) easily understandable; (iii) analysis is critical and good. OUTREACH has been continuously using participatory methods as they have obsolute conviction and belief in the methods.

The staff of OUTREACH had an informal meeting with a group of villagers consisting of 7 men and 6 women in Usthanapalli village. The subject of discussion was problems of the village. The technique employed was Venn diagram. Normally, Venn diagram was applied to study the relationship using circles of different types. OUTREACH improvised this method by using 'squares' of different types for ascertaining and prioritizing problems. (See Exhibit 1(a)

The top most problem as prioritized by the participants is 'lack of road and bridge'. They decided that this problem has to be solved by the local panchayat. Hence, they took up the next important viz., 'Soil erosion' and decided to probe further using the tool 'trend change'.

The trend change enabled the farmers to ascertain the problems associated with soil erosion like depletion of soil, shift in cropping pattern, land availability for cultivation, degradation of forest and other related issues. (See Exhibit 2)

EXHIBIT 1 A : Venn Diagram Usthanapalli

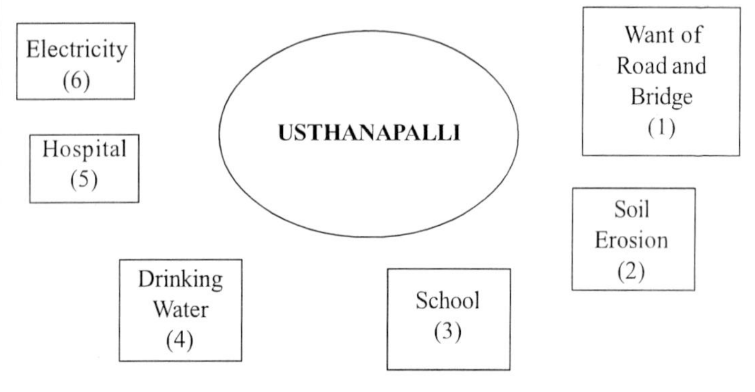

Note: Big square represents top most problems. Small square indicates least problem

The trend change revealed the following:

- top soil has been depleted over a period of 20 years resulting in fall in soil depth;
- flooding has increased;
- tree coverage in the forest has gone down; and
- the extent of land available for agriculture has also gone down (due to expose of rocks).

The participants, along with the staff, discussed all these problems in detail and also the steps to be taken to conserve soil. They decided to use *Seasonal calendar,* as it would help them to plan the items of work that should be taken up to solve the problem. (See Exhibit 3) In the Seasonal calendar, they depicted rainfall, agricultural operations, availability of labour and soil conservation measures to be taken up at different periods of time.

The three exercises helped the people to identify their problems, the seriousness of the problems and activities to be undertaken in order to solve the problems. This was followed by *role and responsibility allocation exercise through focused group*

EXHIBIT 2: Trends

Usthanapalli

Variables	*1978*		*1988*		*1998*	
Agriculture yield						
Ragi	00000	5	00	2	00	2
Paddy		0	000	3	00000	5
Soil depth	00000 00000	10	00000	5	000	3
Flood (run off)	00000 00	7	00000	5	00000 00000	10
Tree species in the forest	00000 00000	10	00000 0	6		0
Land availability for cultivation	00000	5	00000	5	000	3
Population	000	3	0000	4	00000 000	8
Fertilizer Organic manure	00000 00000	10	00000 000	8	0000	4
Chemical fertilizer	0	0	0	0	000	3
Livestock Cows/Buffalo	00000 0	6	00000 00000	10	00000	5
Goat	00000 00000	10	00000	5	00000	5
Sheep		0		0	00000 00000	10
Cattle diseases	00	2	00	2	00000	5
Crop diseases	00	2	0000	4	00000 00000	10
Literacy						
Men	0	1	00	2	0000	4
Women	0	1	000	3	0000	4

Note: Stones have been used to study the trend. Ten stones indicate the HIGHEST prevalence and one stone is the LOWEST.

EXHIBIT 3: Usthanapalli: Seasonal Calendar

Months Details	*January*	*February*	*March*	*April*	*May*	*June*	*July*	*August*	*September*	*October*	*November*	*December*
Rainfall	–	–	●	●●	●●	●●●●	●●●	●●●●●	●●●●●●●	●●●●●●●●●	●●●●●●	●●●●
Agricultural operations	Harvesting of tablab and redgram	Thresing of ragi	Thresing of ragi	Ploughing	Sowing of gingelly	Sowing of groundnut	•Harvesting of gingelly •Sowing of ragi and samai Weading operations in groundnut field	Weeding operation in ragi field	•Harvesting of ground nut •Sowing of horse gram	Harvesting of ground nut •Sowing of horse gram	Harvesting of ragi	Harvesting of sorghum
Labour availability	30	20	25	40	40	20	10	20	—	—	—	—
Soil Conservation activties proposed												
Land cleaning	√	√										
Bunding	√	√										
Conversion												
Drain		√										
Check-dam		√	√									
Pitting				√	√							

discussion on how to implement the work related to soil conservation. This exercise helped them to allocate the responsibilities for executing the work. The participants from the village have agreed to take up the responsibility of cleaning the land, bunding and farming diversion drain. The OUTREACH has agreed to construct a check dam and assist the participants in planting tree seedlings wherever there has been soil erosion.

CASE II

Participatory Methods facilitated Action

This case is about a Self-Help Group, which established micro-enterprise through participatory process. In the process of establishing the micro-enterprise, the group used certain participatory methods, which enabled them to promote such an enterprise.

Om Sakthi Mahalir Mandram (Womens' Self-Help Group) was promoted by OUTREACH. The group was regularly saving money and providing credit to the needy among its members. But then, the ultimate aim of the group is not mere savings and credit, but establishment of micro-enterprises for promoting the economic conditions of the women. The staff of OUTREACH, in one of the monthly meetings of the association, initiated a dialogue on the feasibility of starting an income generation project for the group. The members of the group discussed various options like running a ration shop, starting a provision shop, or a rice shop, establishing an agro-service centre and so on. They discussed about the profitability and feasibility of different options and finally settled on starting an agro-service centre. However, they decided to probe further on the viability of the agro-service centre. The technique they adopted first was *mobility map* (see Exhibit 4) as it was likely to help them to find answers to key questions like: Where do farmers buy agro-inputs? What are the different types of agro-inputs they buy? How much distance do they need to travel? What is the transport cost? What are the difficulties? Where could the proposed centre be located? The mobility map helped them to know the location of existing Agro-service centres (private and cooperative), the villages served by these centre, total households

EXHIBIT 4: Nandinangam: Resource Location and Mobility Map

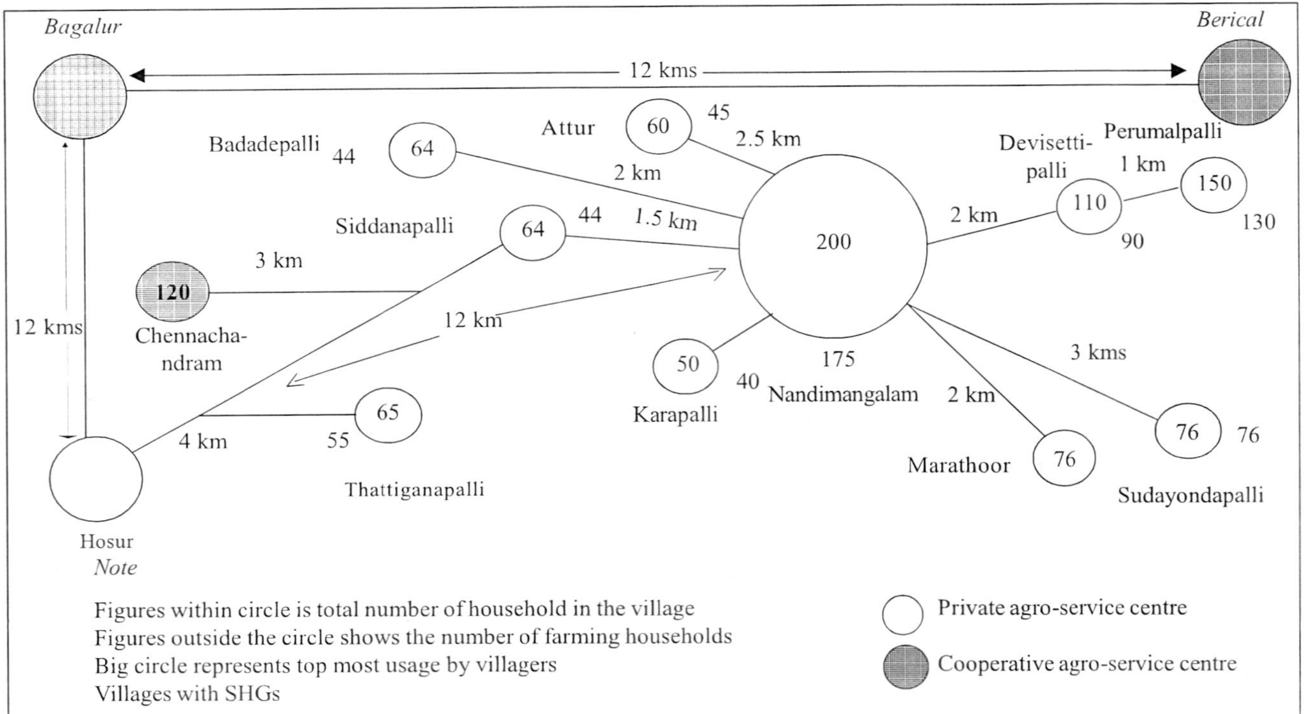

and farming households in those villages, presence of self-help groups in those villages, distance between different villages etc. These information helped them to find out the potential customers, feasibility of establishing link with other self-help groups and the location of the proposed centre. They also discussed the advantages and limitations of establishing the centre at their village Nandimangalam.

Having completed the preliminary discussions on starting an agro-service centre, the group decided to ascertain constraints and opportunities in promoting the centre. They employed *SWOT analysis* through *Guided Dialogue Technique* for this purpose. (See Exhibit 5)

The members of the Sangam realized, through SWOT analysis, that strengths and opportunities for starting the agro-service centre outweigh the weaknesses and threats. Hence, they decided to go ahead with the organization of the centre. They discussed the objectives of the proposed centre thoroughly and finalized. They are profit, skill development in micro-enterprise and service to the community.

The members of the Sangam then had a detailed discussion with the CLA. The results of the appraisal along with visuals were presented. A budget was prepared for Rs.49,000. Contribution from CLA and OUTREACH was discussed and finalized. OUTREACH agreed to provide an interest free loan of Rs.39,000 to CLA for lending to SHGs for a period of three years. Members of the Sangam contributed Rs.10,000. In order to operate the affairs, Purchase committee and Sales committees were constituted.

The Agro Service Centre was established with 19 women members. They run it by "one-member-one-day" basis. The centre opened during 6.00-9.00 a.m. and 5.00-8.00 p.m. (the timing was fixed based on the farmers' interest). Almost one year is over. The annual turnover is Rs.2.5 lakhs and the centre has earned a net profit of Rs.11,000/- after paying an amount of R.10,000/- to OUTREACH towards loan repayment. Earlier, the centre was operating in a rented building. Now, it is operating in a building constructed by Sangam members who raised loan from CLA through Sangam.

EXHIBIT 5: SWOT Analysis on Proposed Agro-Service Centre

Strengths	*Problems/Weaknesses*	*Opportunities*	*Threats*
Availability/supply of agricultural inputs on time	Overstocking of inputs	Possibility of forming different committees like purchase committee	Lack of administrative ability among the sangam members
Generation of profit for sangam	No knowledge in maintenance of accounts	Timely supply of required inputs	Competition from private traders
Development of business skills among the members	Paucity of working capital	Training in accounts keeping to members	Insistence on credit transaction
Linkage with government (Agricultural Department etc.)	Obtaining license	Problems related to obtaining license, payment of tax may be solved as the sangam is the member of IFAD	Failure of rainfall
Reduction in unnecessary expenditure to farmers	Non-cooperation from the people	Possibility of hiring a building	Excessive rainfall
Creation of good name to sangam	Lack of entrepreneurial skills	Motivating other members of the sangam for purchase of inputs from center through cluster level association (CLA)	
Provision of service to adjoining village on time	Insisting credit transactions by the people/public	No credit transaction. But SHGs can lend their N.P. number CLA and OUTREACH for working capital assistance	
		Ineffective functioning of Co-operative Agro-Service Centre	

CASE III

They can do it
(Based on Direct observation)

This case reports on how even an illiterate women's group could conduct the meeting in an orderly manner, and minute the meetings with meticulous care on the spot.

The venue was a small room under an overhead tank of the village. The time was around 7.00 p.m. Totally, 17 members of Om Sakthi Mahalir Mandram and a few staff from OUTREACH were present. Barring the book-writer, no member of the Sangam had formal education. It was a weekly meeting of the Sangam.

The members present selected a moderator for the meeting. The role of the moderator is to facilitate the deliberations of the meeting. The meeting started with a group prayer. The moderator requested the members present to propose the agenda for the meeting. The agenda as proposed and finalized are:

- Sharing of experience by the members who attended a training programme at OUTREACH.
- Preparation of Agriculture Plan for the current year
- Repayment of IFAD loan (mobilized from Indian Bank)
- Problems in running the Agro- Service Centre.

Every one of the members present actively participated in the deliberations. The proceedings and resolutions passed were meticulously recorded in the Minutes book. They were read out and the members approved by affixing their signature in the book.

The occasion was also used to collect the subscription and recover the loans issued to members. All the transactions were recorded in the books of accounts on the spot.

The members of the group then discussed the problems in running the Agro-Service Centre. A member of the Sangam facilitated. No one from OUTREACH intervened. The facilitator organized the exercise with ease and in a very systematic manner. Problems, as identified by the participants were prioritized using *card-sorting* method. The problems as identified and prioritized are:

- — committee members are not punctual in making purchase
- — transport
- — no space for storing cattle feed
- — demand for credit sales
- — members as per rotation do not attend sales duty

They discussed the steps to be taken to overcome the problems. The members are quite aware of the method. All of them actively participated. We could observe a system, science and an art in their approach and way of conducting meeting.

CASE IV

Participation empowers

This case shows how a shy woman, within a short period of time, could become a President of the Self Help Group and represent CLA. The base for her rise in the social status is attributed to her ability to manage a situation using participatory methods. She has evaluated herself among a group of women using an exercise called *'Self-evaluation'*.

Lalitha, 28 years old, married, comes from an agriculture family. She has passed eighth standard. Persuaded by Jeyamma, she joined the Sangam 18 months back. She felt very shy and said "I did not know how to interact and mingle with people when I joined the Sangam. But now I can speak freely with confidence". She attributed this to her participation in many meetings and use of participatory methods. In one of the meetings, while explaining the significance and merits of participation, she evaluated her progress after joining the Sangam. (See Exhibit 6)

The practice of that exercise, she feels, is a sort of self-introspection. It has enabled her to analyze how she has improved and what factors have been responsible for such progress. Such an exercise in a group enables others to evaluate themselves. Others too, try to improve their position. Even a dull member, she opines, becomes so active.

EXHIBIT 6: Self Evaluation of Lalitha—After Joining Sanga –Through Participatory Method

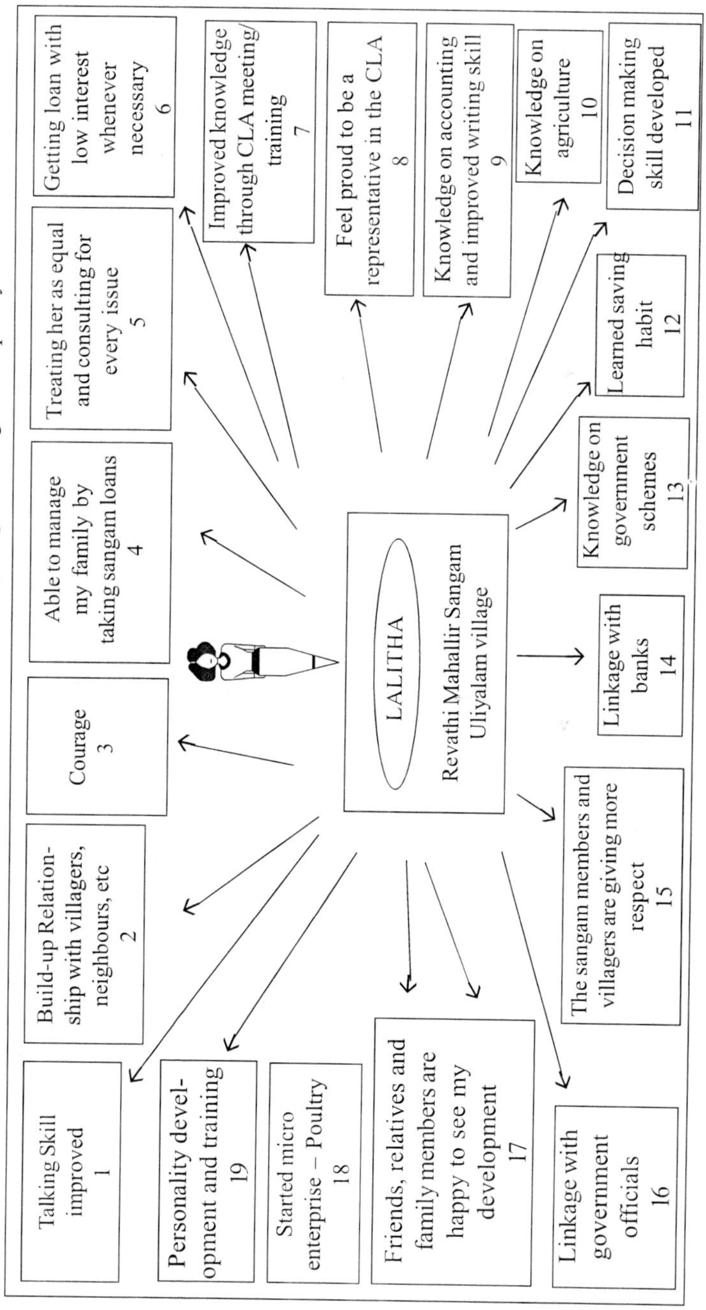

Lalitha is now President of the Sangam and has been elected as representative to Cluster Level association (CLA).

CASE V

Sequencing of Methods and Action

Participatory methods not only help in initiating action but also help the participants to reflect critically on the outcome of the action. Such a critical reflection helps the people to point out where they have erred and how such errors should be rectified.

The Ganga Mahalir Mandram, a self-help group, promoted by OUTREACH in one of its weekly meetings worked out the member-wise loan budget (credit requirement) for the ensuing year using *Seasonal Calendar.* The budget was prepared under four heads viz., agriculture, education, festival and marriage (See Exhibit 7). The members asked for loan from the SHG. But then a perusal of the records revealed that many members of the group did not repay the loan received earlier from the SHG. Therefore the members, facilitated by OUTREACH, analyzed the reasons for non-repayment of loan by using Chappathi Diagram (See Exhibit 8). The exercise revealed the following reasons:

— redemption of prior debt to middlemen,
— want of place for storage, and
— paucity of place for drying the groundnut.

These three factors, the members felt, led to low staying/ bargaining power resulting in selling of groundnut immediately after harvest which ultimately fetched a low price for the groundnut. This has ultimately led to non-repayment of loan to the SHG.

In the course of discussion, the members keenly felt the problem of middlemen. Along with facilitator, the members therefore, wanted to analyze the role of middlemen at different stages of the cultivation of groundnut and marketing using *Seasonal Calendar*. (See Exhibit 8A)

The seasonal calendar enables the members to know how middleman traps them and also to identify post-harvest credit

EXHIBIT 7: Major Expenditure in a Year Sudagondapalli Ganga Mahalir Sangam

Months[1] Details	*January*	*February*	*March*	*April*	*May*	*June*	*July*	*August*	*September*	*October*	*November*	*December*	*Total*
1	2	3	4	5	6	7	8	9	10	11	12	13	14
Lakkamma	500*		5000*			1000√ 2000^{+}	1000√		1000* 2000√	3000*			15,500
Jeyamma	1500*		1000*		3000√	2000√ 1000^{+}	1000√		500* 2000√	4000*			16,000
Sinnathayamma	500*		2000*		1000√	1000√ 1000^{+}	500√		500*	2000*			8,500
Nagamma	500*				1000√	2000√	1000√		1000* 1500√	1000*			8,000
Lakummama			2000*			1000√	1000√		2500√	1500*			8,000
Thikkamma	500*		3000*		1000√	1000√ 500*			500^{+} 1500√	2000*			10,000
Kuramma	500*		4000*	60,000•	1000√	2000√ 500^{+}	1000√		1000√	2000*			72,000
Sinnamma	500*		4000*		2000√	2000√ 1000^{+}	1000√		500* 2000√	3000*			16,000

Contd......

Exhibit 7: *(contd.)*

1	2	3	4	5	6	7	8	9	10	11	12	13	14
Lashma-mma	500^{*}		2000^{*}		1500√	1000√ 1000^{+}	500√		1500√	1500^{*}			9,500
Rajamma	500^{*}		1500^{*}			1000√ 500^{+}	500√		500√	1500^{*}			6,500
Naraya-namma	500^{*}									500^{*}			1,000
Kemba-mma	500^{*}		2000^{*}						500^{*}	1000^{*}			4,000
Nanjga-mma	500^{*}			30,000•					1000^{*}				31,000
Pillamma	500^{*}		1000^{*}						500^{*}	500^{*}			2,500

x Festivals + Education

√ Agriculture • Marriage

EXHIBIT 8: Reasons for the immediate sale of groundnut after harvest

Village: Sudagondapalli

Repayment of debts (Loan got for groundnut from middlemen) | No place for storage | No place for drying the groundnut

* More number of 'squares' indicate high intensity of the problem.

requirements of the members. It also made them to understand the need to distribute credit by the SHG in instalments. It has further helped them to think in terms of 'value addition' to groundnut by undertaking processing activity.

CASE VI

Self-Monitoring

Participatory methods can be used to monitor the various activities undertaken in a project. OUTREACH extensively used *scoring* method to monitor the functioning of the self-help groups. It is not monitoring by OUTREACH using externally evolved criteria. It is a sort of self-monitoring based on people's criteria. Such a monitoring results in immediate action.

The various groups functioning with OUTREACH have monitored their performance by using scoring methods. To start with, the group finalizes the criteria based on which the performance of group has to be evaluated. Each criterion was assigned a score of ten. The members as a group have to evaluate the working of their group by assigning a score. The criteria evolved and the scores assigned are presented in Exhibit 9.

The grading, as done by the group, has shown that the group

EXHIBIT 8A: Groundnut Cropping Pattern

Sudagondapalli

Months / *Farm Activities*	*January*	*February*	*March*	*April*	*May*	*June*	*July*	*August*	*September*	*October*	*November*	*December*	*Total*
Land preparation				✓	✓								
Groundnut peeling				✓									
Seed sowing						✓							
Weeding							✓						
Application of fertilizers							✓						
Middle men (Cultivation)				❦									
Harvest								✓	✓				
Drying groundnut										✓	✓		
Middlemen (Marketing)								❦	❦	❦	❦		

EXHIBIT 9: Self-Monitoring of Kamakshi Mahalir Sangam

Kothapalli village

Criteria	*Score of 0-10*
Regular meeting	OOOOOO (6)
Attendance (100%)	OOOOO (5)
Punctuality	OOO (3)
Regular savings	OOOOOO (6)
Discussion and decision making	OOOO (4)
Rules and regulation	OOOO (4)
Loan repayment	OOO (3)
Book - keeping	OOOOO (5)
Bank remittance	OOOOOO (6)
Attending meetings and training	OOOO (4)
Unity	OOOOO (5)
Total	**51**

has scored 51/110. The areas in which the group was found to be weak are:

- group loan repayment
- attendance (meetings and trainings)
- discussion and decision-making
- punctuality
- unity

The group discussed the 'grey' areas in detail and worked out ways and means to strengthen the group. One important suggestion emerged out of the group meeting was to request the OUTREACH to organize training in the areas in which they are weak.

CASE VII

Participatory Training

OUTREACH introduced participatory methods in all its training programmes. Using participatory methods in the training programme makes the training absolutely participatory and in fact, the trainees' role is much greater than the trainers' role.

The OUTREACH organized a one-day training in leadership development for office bearers of the self-help groups it had promoted. A total of 18 women participated in it. Participatory training methods were used to impart training. The purpose of the training is to develop leadership qualities among the women. The family should provide a congenial and enabling environment for the women members to take up leadership responsibilities. Keeping this in mind, the facilitator of the training programme initiated a discussion on the bad qualities in the family.

The process began with the participants divided into two groups. Each group was asked to discuss the bad qualities as found in their families. The technique of *'brainstorming'* was followed. Following the group discussion, a plenary presentation was arranged. Each group presented and shared their reports on the same issue. Questions were raised; clarifications were sought, consensus was arrived at. A list of bad qualities normally found was arrived at. The members of the group actively participated and the trainer's role was very minimal.

After the plenary presentation, the participants were again divided into two groups. The purpose was to prioritize the bad qualities. One group was asked to rank or prioritize the bad qualities using 100 seeds scoring method while the second group was also asked to rank the bad qualities by using Chappathi Diagram.

Under the 100 seeds scoring method, the group was given 100 seeds. The group has to use seeds for assigning scores to the bad qualities identified. The 'worst' quality will be assigned the highest score, the next 'worse' quality will be assigned next high score; .... and the 'bad' quality will be given the least score. (See Exhibit 10)

EXHIBIT 10: Ranking of Bad qualities in a family
(100 seeds scoring method)

Bad qualities of family members	*Score*	
Fighting with each other without understanding	**********	(10)
Unnecessary expenditure	*****	(5)
Anger	*********	(9)
Lack of freedom to women	*****	(5)
Lack of sharing responsibility	******	(6)
Alcoholism	**********	(10)
Laziness	****	(4)
Unsystematic way of doing things	******	(6)
Jealousy	****	(4)
Fighting with others without any reason	******	(6)
Lack of unity	*****	(5)
Lack of affection	***	(3)
Lack of respect to non-bread winner	**	(2)
Not interested in development	**	(2)
Not interested in education	****	(4)
Want of self-confidence	*****	(5)
Arrogance	**	(2)
Carelessness	***	(3)
Stealing	**	(2)
Abetting dowry	***	(3)
Preventing girl child from attending school	**	(2)
Lack of sympathy	**	(2)

EXHIBIT 11: Ranking of bad qualities in a family – Chappathi Diagram

Bad qualities	*Intensity of problem*
Fighting with each other without understanding	
Unnecessary expenditure	
Anger	
Lack of freedom to women	
Lack of sharing responsibility	
Alcoholism	
Laziness	
Unsystematic way of doing things	
Jealousy	
Fighting with others without any reason	
Lack of unity	
Lack of affection	
Lack of respect to non-bread winner	
Not interested in development	
Not interested in education	
Want of self-confidence	
Arrogance	
Carelessness	
Stealing	
Abetting dowry	
Preventing girl child from attending school	
Lack of sympathy	

Under Chappathi Diagram, the worst quality will be represented by bigger rectangle card; the next one will be given next bigger rectangle card and so on (see Exhibit 11).

After discussion, the groups made a plenary presentation and compared their findings. Steps to be taken to do away with bad qualities were also discussed at length.

At the end of the programme, the facilitator invited feedback from the participants. Everyone gave his or her opinion on the training programme. Prominent among them were:

- an in-depth analysis of the issues could be attempted as all families were involved.
- such an exercise had an everlasting impression.
- the methods and tools used were quite simple and easy to practice. They were quite effective for unlettered people.
- problems were not confined to quite a few families but common to all families.
- the analysis also provided a few avenues in tackling the bad qualities in a family.
- all the members could participate and the degree of participation was high.

Innovations

Chambers says, "it is more important for each person and each group to invent, adopt their own approach, methods and sequences and combinations than to adopt a ready-made manual or model. He adds, "let a thousand flowers bloom and let them be flowers which bloom better and better, and spread their seeds". True to this spirit, OUTREACH has experimented with innovative and improvised methods. A few such innovations are presented here. *Impact Diagram (Ladder)*: OUTREACH used Ladder scale (1-10) to assess the impact of Joint Forest Management. This is very simple exercise through which the project team could easily assess the impact of the programme. For instance, the project team assessed the level of awareness among the people after the introduction of Joint Forest Management. Similarly, the team also assessed the availability of raw materials after the introduction of JFM (See Exhibit 12). The people using the ladder could also predict future.

EXHIBIT 12: Impact Diagram – Ladder Scale

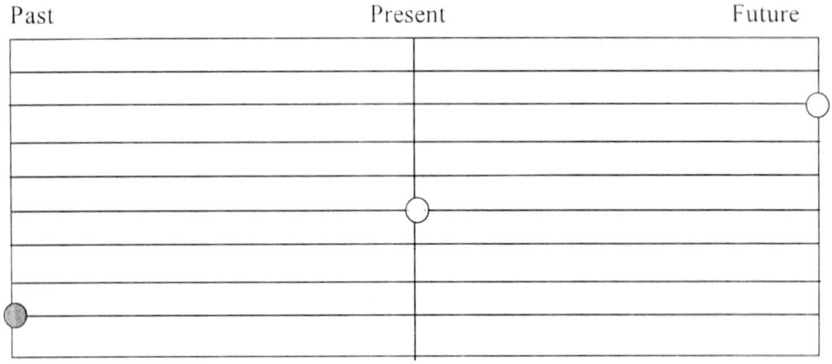

Level of awareness after the introduction of JFM

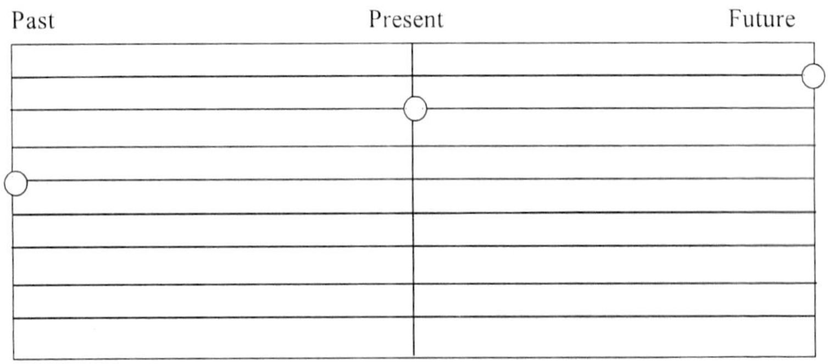

Availability of raw material after the introduction of JFM

EXHIBIT 13: Innovation – Example Assessing the role of VFC members and other users

Activities	*Land owners*	*Landless*	*Basket makers*
Micro plan involvement	○○	○○	○
Representation in VFC	○○○	○○	○○
Participation in meeting	○○	○○	—
Participation in training	○○	○○	—
Participation in implementation	○○	○○	○
Management of forest	○○	○○	○○
Involvement in species selected	○○○	○○	○

Note: Three seeds are the highest and one seed is the lowest.

EXHIBIT 14: 100 Seeds Scoring : Community Choice of Species

Community choice of species	*Seeds*	*Species planted under JFM*	*Seeds*
Sampige	5	Sampige	6
Saguvani	13	Nilgiri	10
Beete	12	Accasia	7
Neem	8	Beete	4
Tamarind	12	Neem	7
Hebbalase	6	Tamarind	8
Taale	4	Halasu	6
Halana	7	Nerale	6
Mavu	11	Gobbaradagida	7
Nelli	2	Buguri	5
Basovan pada	4	Nellie	3
Karijali	11	Basovan pada	6
Nerale	5	Karijali Nugge Hippe Apla	11 3 5 6
Total	100	Total	100

Scoring: This is yet another simple method employed by OUTREACH to assess the participation of Village Forest Committee (VFC) members and other users in various activities related to forest management. This would help the members of VFC to assess their role and also to find out ways and means of strengthening their role. This method is facilitated largely by using seeds. (See Exhibit 13)

100 Seeds Scoring: This exercise is being extensively used by OUTREACH in all their PRA related exercises. For example, they have used this method to ascertain the community choice of species and how it differs from species planted under JFM. (See Exhibit 14)

EXHIBIT 15: Anxiety map through family counselling technique

Activities in dairy enterprise	*Father-in-law*		*Mother-in-law*		*Son-in-law*		*Daughter*	
	Present	*Future*	*Present*	*Future*	*Present*	*Future*	*Present*	*Future*
1	2	3	4	5	6	7	8	9
Buying fodder	○○	●● ●●	—	—	○○○ ○○○ ○○	●●● ●●●	—	—
Milching	—	—	—	—	○○ ○○	●●● ●	○○ ○○	●●● ●●
Collecting grains	○○○ ○○○	●	—	—	○	●●●	○○ ○	●●● ●●●
Bathing	○○○	●	—	—	○○○ ○○	●●● ●●● ●	○○	●●
Cleaning shed	—	—	○○○	●●●	—	—	○○ ○○ ○○ ○	●●● ●●● ●
Taking milk to dairy	○○	●●	—	—	○○○	●●●	○○ ○○	●●● ●●

Contd.....

Exhibit 15: *(Contd.)*

1	2	3	4	5	6	7	8	9
Collecting money	○	●	—	—	○○○	●●● ●	○○ ○○	●●● ●●
Bringing Doctor	○○ ○○	●●●	—	—	○○○ ○○○	●●● ●●●	—	—
Feeding	○○○ ○○	●●	—	—	○	●●● ●●	○○ ○○	●●●
Decision making	○○○ ○○	●	○	●	○○○	●●●● ●●●●	○	●●●● ●●●●

Process

This exercise is related to a family

○ with one animal – the sharing of work and responsibilities.

● with an additional animal what would be the sharing of work and responsibilities.

For each activity, 10 stones are given for assessment of work sharing (present and future).

Anxiety map through family counselling techniques: This is an improved version of matrix scoring. This exercise is related to a family. This is done to know what would be the changed role of different members of family if an additional activity were introduced. OUTREACH had tried this exercise in a family to know how the responsibilities of the members would be shared if one more milch animal were added to the existing number of milch animals. The exercise and the learning experience are presented in exhibit 15.

Inferences from the Cases

The ultimate aim of the participatory methods, as claimed by the propounders, is to empower the people especially the marginalised. It is not just empowerment to articulate; but releasing the poor from the shackles of dependency and leading them to independency. The process of empowerment consists of three stages.

Dependency -> interdependency -> independency

OUTREACH has largely succeeded in making the marginalized, especially the women, to reach the stage of independency. It is a long drawn process. Yet they have made it. The cases presented has revealed that participatory methods

- could enable the illiterate and the marginalised to articulate and act.
- build capacity among the weaker sections especially in the areas of appraisal, planning, implementing and monitoring and evaluation of programmes and activities.
- help develop leadership qualities in the participants.
- help in bringing down the dependency culture; enable the people to realize the importance of self-development.
- help upgrade the analytical skills of participants.
- develop the participants as resource persons.
- develop decision-making capacity among the participants.
- create confidence among the participants-the confidence of "we-can-also-do-it".

SPEECH

Born in the year 1987, SPEECH has incorporated PRA as a prime strategy in its development work. Its ultimate aim is to empower the people. Its development strategy has five stages. They are participatory stage, organizing stage, experimental stage, extension stage, and phasing out stage. Right from the initial stage, people are facilitated to participate. Of course, in the initial stage, the role of SPEECH in development process is on the high side. But as the project passes through different stages, people slowly gain confidence and take over the different activities. At the phasing out stage, people's participation reaches the maximum (95%) and SPEECH's role is nominal and that of facilitating (See Exhibit 16).

Exposed to the concept of PRA in 1989, SPEECH has applied PRA in watershed management, tank rehabilitation/ modernization, local resource management, savings and credit, child labour, education, health, bio-intensive garden, micro-enterprises, conflict resolution, monitoring and evaluation and research. SPEECH with its vast experience in PRA offers training to a variety of audiences like NGOs, Government departments and universities. It is partly responsible for the quick spread of PRA among NGOs in Tamil Nadu.

The participatory methods used by SPEECH are mapping (social, resource, vision, thematic), seasonality, wealth ranking, time line, trend change, matrix (problem, preference, pair-wise), transect, access and control, linkages, and Venn diagram. The most frequently used methods are social and land resource map, seasonality, Venn diagram, linkages, time line and trend change. The reason is that they help in collecting the basic data. Of course, the use of methods vary in relation to nature of the project.

CASE I

PRA in Project Cycle: Paraikulam Experience

Paraikulam is a tiny, remote hamlet located at a distance of 10 kilometers from Tiruchuli. The inhabitants are mostly landless, marginal and small farmers. They are economically backward. SPEECH has rightly chosen this village for its development

EXHIBIT 16: Development Strategy of SPEECH

	Participatory Stage I	Organising Stage II	Experimental Stage III	Extension Stage IV	Phasing out Stage V	
OUTSIDERS (NGOs/Experts)	95 % Basic information Contact with key informants Objective formulation Catalyzing Rapport building Identifying the weaker (PRA) Appropriate approach Need Assessment Village dynamics Area identification Opportunities	Prioritizing problems Problem solving process Govt. contacts worker Efficiency building Self-confidence building Identification of leaders Leadership training Information of people Forums Addressing issues Govt. scheme awareness	Capability building Problem solving process Local resource mobilization Govt. representation Systems review, monitor, Evaluate or follow up Encourage federating Reports Legal formalities	Enable leader's emergence Management Training (SLC) Project coordination	95%	INSIDERS (LOCAL COMMUNITY/SANGAS) POOR
LOCAL	5% Problem listing Opportunity sharing Cooperation 80%	20% Problem analysis Deciding Approach Awareness/Learning Self-confidence Clarity in thinking Justification of issues Ideas into action Team building 60% Prioritising projects Financial assistance	40% Team sprit & work Planning possible solutions Identification of projects Work out Action plans Reports 20% FACILITATION Strengthening sangams Establishing contact	80% Bargaining power increase People's project cycle Problem solving capacity Credit management Legal status of federation Direct reporting Contact (External) Beneficiary selection Trying govt. resources Government representation Follow-up and evaluation 5% Handing over responsibilities Consultancy and counselling	ON-GOING Reflection on different projects Facilitation Control Political Economical Health Government schemes Seeking replicable models Federating at block level Handling training programmes Flexibility General evaluation	OUTSIDERS

'O' ← PROCESS OF EMPOWERMENT → Z

intervention. It started non-formal education centre in the village. The purpose is to eradicate illiteracy among the people and to create awareness on various government schemes. At this stage, the SPEECH adopted rapport-building methods like drama, folk songs, night halts, do-it-yourself, etc. This has helped in establishing rapport with the community. The non-formal education helped the people to gain confidence, which in turn helped them to seek government intervention in solving long-standing problems like streetlight and drinking water with the support of SPEECH. All these created excellent rapport between the people and the SPEECH. Thus, a base has been created for initiating other PRA exercises in the village.

The first step taken by the SPEECH was to assess the situation in the village. It used social mapping to gather background data about the village. There was some initial hesitation; but later, villagers enthusiastically participated. Initially, before drawing the map, the participants told that there are 90 households in the village. But mapping helped them to arrive at the exact number of houses in the village.

After completing the social map, the villagers drew the resource map indicating the extent of land owned, type of soil, crops cultivated, and irrigation facilities. On completion of the map, the participants realized that a vast stretch of the land remain barren and uncultivated. The villagers wanted to show to the participating staff of SPEECH, the conditions of barren land. So they took them for a 'transect' (group walk), which enabled both the staff of SPEECH and villagers to assess the conditions of the barren lands which was once cultivable. Transect helped the participants to ascertain the problems and opportunities available to regenerate the land. The team, after assessing the situation, wanted to ascertain the reasons for degradation of land. Hence they decided to use 'trend change' and 'time line'. This broadly helped them to understand the 'history' of the barren land. The preliminary appraisal and analysis of the barren land, through various participatory methods, made the people to realize the need to regenerate the land. Hence, under the guidance of the staff of SPEECH, the participants from the village started preparing work plan to regenerate the land. The people discussed and decided on where the check-dam has to be constructed, where the bunds have to be raised, where the grass and trees have to be planted to conserve the soil, where the gully check are to be constructed and where the well is to be dug. All these details are depicted on a map, which paved the way for active participation of all the people in the preparation of action plan. The participants, along with the staff of the SPEECH, also discussed and finalized the micro-plan, which include physical and financial budget and work calendar. The contribution from the people was fixed at 40 per cent of the total cost of the project.

The people implemented the project, which was prepared by the villagers under the guidance of SPEECH. People were made responsible for monitoring the project. The project has regenerated the barren land and the people in the village started earning income from the land, which was once barren and uncultivable. The participants, using participatory resource map and trend change, evaluated the project. Participatory resource map and trend change drawn at the time of appraisal was used as the base for assessing the impact of the project.

A very interesting aspect of the experience is its spin off effect. The people of the neighbouring village (Kottam) were observing the developments in Paraikulam. Impressed by the activities of the people, the people at Kottam who had more or less similar problem, organized a PRA exercise with the help of Paraikulam people and prepared the micro-plan (65 acres) using participatory methods. The plan was submitted to SPEECH for financial and consultancy support. SPEECH carefully considered the proposal and provided 20 per cent of the total project cost.

CASE II

PRA in Rehabilitating Minor Irrigation Tanks

This case is on SPEECH's collaborative effort with the government in rehabilitating a tank, using participatory methods. The government, which has been following top-down approach, has realized the significance of local knowledge, bottom-up approach and the benefits of working with the people.

The state of Tamil Nadu is characterized by numerous small

irrigation tanks. These tanks most of them rain fed, covers a large area of apparent wasteland and generally, have the capacity to irrigate 20-40 hectars of farmland. During the dry season, it is very difficult to distinguish severely deteriorated tank-bed land from non-tank land. Lack of maintenance, accumulated siltation, and the disappearance of traditional tank management practices resulted in a widespread and significant reduction in tank storage capacity. Consequently, the agricultural output and employment opportunities had also declined. In order to combat this, the Agricultural Engineering Department of the Government of Tamil Nadu undertook a Standardization of Minor Irrigation Tanks (SMIT) programme. This programme aimed at restoring the tank capacity and improvement in overall water management.

The SPEECH, in collaboration with Agricultural Engineering Department, used PRA for the first time in SMIT programme in Vadapalai village. The purpose is to rehabilitate the tank using participatory method.

After preliminary discussions on the purpose of the interaction with the people, the people prepared a resource map and model of the tank. A detailed seasonal calendar provided a wealth of information on labour availability, water flow, rainfall, rainfall intensity, frequency of storms and so on. Three transect walks were made covering feeding channels, water spread area, and the tank bund and command areas.

The knowledge that was shared by the farmers through PRA revealed that

- the farmers were able to mark and locate the feeding channel requiring repairs; specific spots requiring stone revetments; the exact location in the irrigation channel where division boxes needed to be constructed and the boundaries of unauthorized irrigated land,
- the maps produced by the farmers were later found to be extremely accurate and realistic when compared to survey map,
- the farmers were able to pinpoint exactly what repairs

and alterations were needed to the infrastructure of the tank, and

- the 20 hectares of registered command area were divided into 80 parcels of land owned by about 65 farmers.

The advantages as claimed by the farmers, Agricultural Engineering Department and SPEECH are:

— use of participatory methodologies based on open attitude and behaviour was found to be more effective than the conventional approach.

— participatory methodologies enabled the farmers and the Agricultural Engineering Department personnel to develop an effective working partnership. Everyone involved was more aware of what was required and the value of each other's contribution to the process.

— collection of data was more rapid and the information obtained was more concise, targeted and relevant to the project.

— the resource maps and models provided an excellent visual record of the tank and the current state of its structures. The maps also provided a good basis for planning the subsequent engineering survey, which saved time and money.

— by enabling farmers effectively and clearly to communicate their detailed knowledge about the local environment and conditions of the tank, the use of PRA undoubtedly reduced the overall time for project completion. The farmers clearly derived a lot of satisfaction from being involved in the planning and decision making process for the rehabilitation of their tank, as demonstrated by their greater enthusiasm for taking part in the actual restoration work.

— the use of PRA was helpful in exploring the areas, which were too technical for farmers. This included the water seepage through the tank-bund, sub-optimal flow resulting from the division of sluice flow, and the excessive irrigation. By bringing these issues into open, they could be discussed and resolved.

— by establishing more equal participation between Agricultural Engineering Department and villagers, the resources were more effectively targeted and completion time of the project reduced. This maximized the benefit of the government investment and helped to bring about more long-time and sustainable benefits to the village community.

CASE III

Participatory methods benefit village Panchayat

This case is presented to show that the practice of participatory methods is more likely to create awareness and action.

SPEECH organized a series of training programmes for the Ward members and the Panchayat Presidents. The training, using participatory methodology, was on various schemes available, access over the schemes, common property resources, access and control over common property resources by the panchayat, and so on.

This training brought a tremendous change in the functioning of the panchayat. Each ward member using participatory methods has appraised the problems in his/her respective village, solutions to problems have been discussed, alternatives have been suggested and the best alternative has been chosen. Budgets have been worked out. A report consisting of all these details have been submitted to the president. The president, in turn, convened the meeting of the ward members, discussed and finalized the proposal and submitted the same to the block development office for funding. The President and ward members normally do the government schemes. The use of participatory methods helps the functionaries of the Panchayat to present a micro plan that represent people's priorities and aspirations.

Innovations

SPEECH is responsible for developing participatory village census. Census maps show social details representing people and household characteristics with local materials such as different seeds, stones and vegetables, on markers such as *bindhi*. This is, in fact, a practice developed by Anusuda and Perumalnaicker of Kethanaickenpatti village near Madurai where SPEECH serves. Under this practice, for each household they will have a card and mark details with symbols on the cards. These have been placed on cards on or the ground on the maps or models to indicate for each household the number of men, women, and children, assets owned, wealth, poverty, the handicapped, immunization status, education and other information. With an informed group or person, a participatory census of a small village has been conducted in less than an hour, and then other information added by 'interviewing the map'.

Inferences

— SPEECH has applied the concept of PRA at different stages of the project. This has enhanced the participation of people. Participatory methods have helped SPEECH to build-up confidence among the partners by slowly and steadily enhancing the role of partners through different stages of the project.

— SPEECH has made PRA as an integral component of the project cycle. It has used relevant, appropriate participatory methods at different stages of the project. It starts with participatory rapport building exercises and ends up with participatory evaluation exercises.

— SPEECH has proved that participatory methods could empower the people. SPEECH has clearly demonstrated that people could analyze their conditions and problems, could find solution to these problems, could execute work and could monitor and evaluate their performance.

— Participatory methods like mapping could be as accurate and realistic as survey map.

— Participatory methods could develop effective partnership among different stakeholders of the project.

— Collection of data through participatory methods could be more rapid; the data could be more accurate, concise and relevant to the project.

— Participatory methods were proved to be less costly and less time-consuming which ultimately result in maximum benefit to the participating villagers.

— SPEECH has proved that institutions like panchayat can prepare micro plan for the panchayat using participatory methods. Panchayats, instead of executing what the government orders them to execute, can meaningfully prepare micro plan integrating the aspirations of the people in the plan using participatory methods.

ASSEFA

ASSEFA known as Association for Sarva Seva Farms is a Non-Governmental Organization. Established in 1969, the organization aims at creating self-reliant rural communities. The organization works throughout India. The case study is related to ASSEFA's Lathur project, which was established in January 1990. The project, with a jurisdiction of 72 villages, aims at holistic development of target population, which include economic and cultural aspects of the life. The activities of the project centre around formation of groups, promotion of thrift and savings, provision of loans to farm and non-farm activities and so on. The strategy is people-centered approach through groups and Gram Sabah. (See Exhibit 17)

EXHIBIT 17: Village Organisations–Linkages

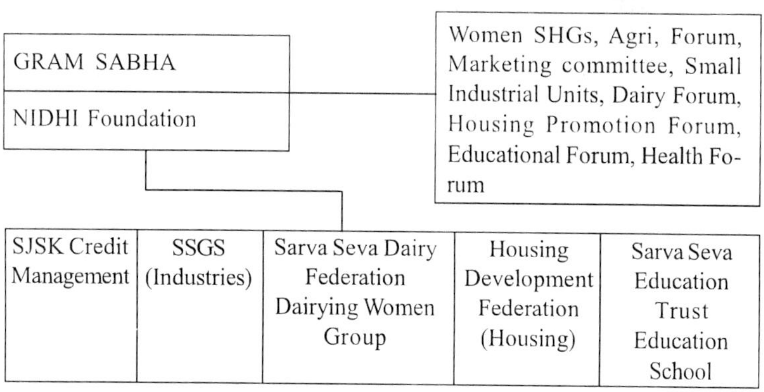

ASSEFA's Lathur project has been using participatory methodology right from their inception especially at the appraisal and planning stage. ASSEFA, as reported by the Project Coordinator, totally relies on participatory methods for appraisal and planning, as there are no other methods as suited and as reliable as PRA. ASSEFA feels that PRA establishes a direct contact with the people. The staff found survey data unreliable and it took too long a period to complete. The areas where PRA applied are: (i) agriculture and animal husbandry; (ii) savings and credit; (iii) women development. The most frequently used methods by ASSEFA are social mapping, resource mapping, wealth ranking and venn diagram. These methods, the staff of the project feel, provide enough information to find out the target group and formulate the project.

CASE I

Project formulation through PRA

Projects emerge from the people. ASSEFA encourages the people to practice PRA for formulating projects as it facilitates direct contact among the stakeholders. This case is presented to show how participatory methods facilitated project formulation.

Projects for promoting the welfare of the partners are formulated in three phases. In the first phase, the Gram Sabha (Village council) using participatory methods, prepares a detailed plan. The staff, at this stage, play a minimum role as facilitators. The villagers are given total freedom to assess the current situation, plan the activities and decide their contribution in the project.

The plan prepared by the villagers is discussed in the second stage. The field staff participate in the discussion and finalize the projects in consultation with the villagers. A final shape is given to the project at this stage.

In the third and final phase, the project team at the headquarters thoroughly discusses the project with reference to the objectives, activities and budgets.

The projects as formulated by the people, by and large, are accepted without much modification. Technical details required, if any, are added.

The organization adopts participatory methods to formulate the project. The methods followed are social mapping, wealth/well being ranking, venn diagram, logical framework analysis and SWOT analysis.

Social mapping is used to know the demographic details of the village and infrastructure available in the village. Wealth/well being rank was employed to identify the target group of the project. Venn diagram is used to know the present status of external linkages and to identify the potential resources for implementing the project effectively.

At the planning stage, "Logical Framework Analysis" is used. Fixing indicators which can easily be achievable with verifiable means is the simple meaning of LFA. The LFA is done with active participation of the people. LFA enables the villagers to define the aim, objectives, responsibilities, and indicator of achievement in more clear and precise terms. The appraisal based on PRA helps them substantially to build up Logical Framework Analysis (See Exhibit 18).

CASE II

Impact Assessment through Social Mapping

This case is presented to show how participatory mapping was used to study the impact of a project.

Keelneerpallam, a tiny village in Cheyyur Taluk, comes under the project area of ASSEFA. ASSEFA assessed the situation of the village using social mapping in January 1996. People actively participated in the exercise. All the basic data (quantitative data) were gathered and shared among the people using social mapping. The people gathered 124 types of quantitative data using locally available materials and verified the data then and there. Data were helpful in formulating the project. The project aims at integrated development targeting on different aspect of life of the people like agriculture, health, sanitation, housing, education, animal husbandry, savings and credit and women development. The target families were fixed using wealth/well-being exercise. The institutions, which provide services and their effectiveness, were analyzed by using venn diagram. Plans were prepared by using LFA. The project was launched in 1996. After 2 years, the people, using participatory mapping, evaluated the project. Base line data collected by the people through participatory methods was the basis. The data and information collected before and after the project was compared

EXHIBIT 18: Logical Framework Analysis

Project title : Socio-Economic Development Programme, Lathur Block (East)

Project description: To improve the socio-economic conditions of the poorest 1127 families in 25 villages of East Lathur

Project Structure	*Indicators of Achievement*	*Means of Verification*	*Risks and Assumptions*
BROAD OBJECTIVES			
To improve the socio-economic condition of the poorest households in 25 villages of East Lathur in a sustainable manner	– The members of 1127 target families will have an annual income of at least 18,000/- rupees per year by 2002	Annual household baseline survey	The absence of a prolonged period of drought and cattle crop disease epidemics
IMMEDIATE OBJECTIVES			
1. To strengthen all GSs and MMs to the level of Self Management	By 2002, all GSs and MMs will (independently of Project facilitation): – maintain updated, accurate accounts – conduct at least one monthly meeting – rotate leadership annually – utilize loans for their intended and make repayments on time	– GS/MM records and registers – GS/MM and Kosh records and registers, Project Design Outlines	– The attitude of the target group continues to be positive towards ASSEFA and other external development agencies – Cooperation and coordination between ASSEFA and other external institutions continues – Village forums are not adversely affected by political affiliations – There is parental support for the education programme
2. To promote community values	– By 2002, all finance required for community development will be raised by the community. – Community issues are discussed at every GS and MM meeting by 2000	– GS/MM records and registers – GS/MM meeting minutes	

EXHIBIT 18: *(Contd.)*

	– 80 % of targeted families will have a representative in either the GS or MM by 2002	– GS/MM records	
3. To promote on and off-farm employment and self-employment opportunity	– At least one adult member from target family will have access to more income generating programme	– Project Design Outline records – Annual household baseline survey	
4. Continue the emphasis on women's empowerment	– 50% of project funds will be directed exclusively to women's programme and training – 75% of women from the target group will have undertaken relevant employment skill training by 2002	– Project records and registers – Project training records	
5. To improve community health standards 6. To improve the performance of school children aged between 5 and 15	– Reduced IMR from 14 per 1000 to 5 per 1000 by 2002 – 95% of targeted village will have clean potable drinking water by 2002 – To reduce the drop out rate of village children from 2.1% to 1% by 2002 – 90 % enrollment of village school children by 2002	– Health animators, monthly reports and registers – Village teachers monthly reports and registers	

EXHIBIT 18: *(Contd.)*

OUTPUT			
1. i) An adequate Revolving fund established in every every village	– Each targeted household will have Rs. 10,000 as share of the village Revolving Fund by 2002	– GS/MM records and registers – Project and Kosh records	– The Target group continues respond to training
ii) GS and MM members will have acquired the necessary skills for self-management	– A self management training every month for all GS and MM members throughout the duration of the project	– Project training records	– External agencies support for training continues
2. i) To raise community consciousness towards community cooperation and welfare	– A minimum of one 'community work' undertaken in every village per month with at least 90% attendance of GS/MM members – All village disputes between members resolved locally through traditional structures without involving the police or law courts – A minimum of two community assets created in every village by 2002	– GS/MM records – Gs/MM records and registers – Project records	
3. i) Targeted families will have obtained the necessary skills to make maximum use of project assistance	– At least 95% loan recovery	– GS/MM Project and Kosh records	
ii) More credit opportunities are available to the target groups	– Every target family will have had at least one loan by 2002	– Project records and registers	

EXHIBIT 18: *(Contd.)*

iii) Farmers in the target group will receive a higher price for their produce iv) Milch animal owners in the target group will receive a higher price for their milk v) Healthier and more productive animals vi) The regeneration of rural artisan trades	– At least a 20% rise in income for paddy farmers – At least a 20% rise in income for milch animal owners – 10 types of rural artisans in the project area by 2002	– Individual and village paddy records – Individual and village milk records – Project records and registers	
4. i) Enhanced awareness of women's position in society ii) Increased opportunities for women to acquire greater responsibilities in the household and community	– Equal gender representation on village functional forums by 1999 – All the willing, eligible women have control over at least one income generating asset by 2002	– Village functional forum records – Individual, MM, Project and Kosh records	
5. i) Every village will have a self managed health programme, comprising an adequately trained local health person and access to treatment for common ailments	– 25 health animators trained in basic health issues by 2002 – one basic medical kit supplied to every village by 2000	Project training records and minutes	

EXHIBIT 18: *(Contd.)*

ii) Increased health awareness and knowledge			
6. Every village will have a self managed night school with a trained local teacher	– 25 night schools operating daily by 1999 – 25 teachers will have received basic teacher training by 2000	Project education records and minutes	
ACTIVITIES		**INPUTS**	**PRECONDITIONS**
1.i) Deployment of adequate funds for programme needs ii) The consolidation of self-management traning			
2.i) Community values awareness programme, using multi-media ii) Village level community works			
7. i) Assessment of staff training needs ii) Every staff member will undergo 4 relevant training sessions a year			

EXHIBIT 19: Impact Assessment

Agriculture

- the extent of land under cultivation has increased from 111 acres to 146 acres.
- the extent of land with assured irrigation has increased from 118 acres to 146 acres.
- number of bore wells has increased from 0 to 21.
- productivity of paddy has shot up from 15 quintals to 20 quintals per acre.
- the net income from paddy has increased from Rs.3000 to Rs.5000.

Animal Husbandry	1996	1998
No. of milch animals		
local breeds	45	30
hybrid	30	51
backyard poultry	4	20
goats	—	10
No. of veterinary camps conducted	—	12
Yield per animal per year (litres)		
cow	360	920
buffalo	450	630
Net income per animal per year (in Rs.)		
cow	2520	4640
buffalo	3600	3240

and the inferences drawn. (See Exhibit 19) They are presented here.

The impact analysis using participatory mapping was done without much of outsiders' assistance.

Such an analysis by the members of the Grama Sabha helps them to understand 'where were they then?' and 'where are they now?' It builds in them a 'confidence' and a sense of achievement.

ASSEFA has also used trend change to assess the impact of its programme on village and also on the women (See Exhibit 20 and 21).

CASE III

Participation Empowers

Promotion of savings and credit through women's association is an important activity of ASSEFA. The credit in turn must result in production-oriented programme, which in turn must lead to increased income to the women. The Lathur project has 25 women groups. All the groups have dairying as a major income generation project. The production of milk therefore was high. Increased production has led to some critical problems like fall in the price, delay in payment, delay in milching etc. They had to sell the milk to the private traders. The representative of the various women's group met and discussed the problems in detail. All of them desired to extricate themselves from the clutches of private milk vendors. Hence, they wanted to start their own marketing organization, which resulted in the birth of Sarva Seva Dairy Farm (SSDF). The SSDF hired a vehicle (21.6.98) for transporting the milk to their milk-chilling centre where the milk was processed. But then they encountered problems with the hired vehicle like high rent and delayed arrival. The Women's group discussed these issues in detail and took a decision to buy a vehicle. The total cost of vehicle was Rs.4.11 lakh. SBI provided a matching credit of Rs.4.11 lakh at an interest rate of 12.24 per cent. The bank gave the loan after gaining confidence in the functioning of the group through its savings.

The women's group managed the vehicle it had purchased. Each woman who supplies milk has to pay a rent of Re.1 per litre of milk. The total income by way of rent works out to Rs.36,000 per month. The expenses like salary of driver and cleaner, diesel, maintenance expenses and other expenses roughly works out to Rs.13,000. The loan repayment per month comes to Rs.14,024. The women's group could easily pay the amount. The members of the various women's groups are totally satisfied with the arrangement.

EXHIBIT 20 : Evaluation of Project Area Integrated Rural Development Project ASSEFA - Sivaganga

	Before intervention of ASSEFA	After intervention of ASSEFA
EDUCATION		
PUBLIC PROPERTY	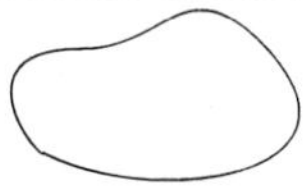	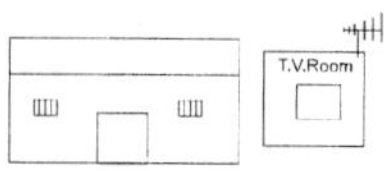
DRINKING WATER	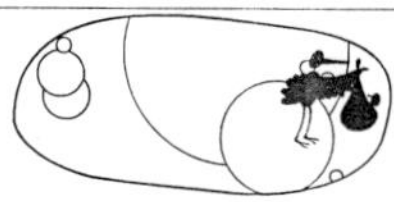	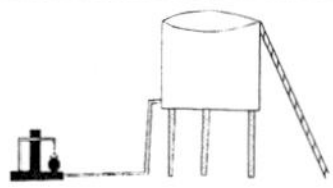
SMOKELESS CHULAH		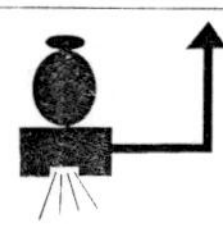
HEALTH STATUS		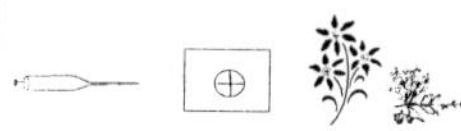
COMMUNITY AND ORGANISATION	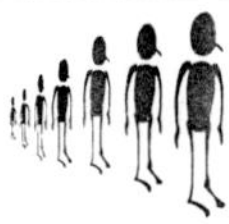	
ECONOMIC STATUS		
KITCHEN GARDENS		
FINANCIAL ASSISTANCE	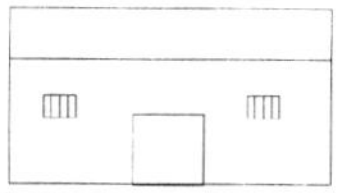	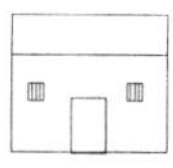 

EXHIBIT 21 : Evaluation of Project Area Integrated Rural Developoment Project (Women status in Easanur village)

	Before intervention of ASSEFA	After intervention of ASSEFA
AGE AT MARRIAGE	Before 18 years	After 22 years
DOWRY		
WIDOW REMARRIAGE	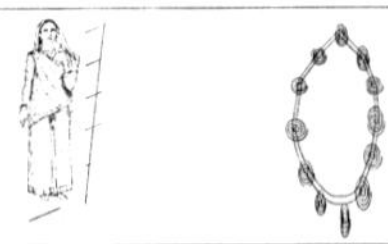	
FAMILY PLANNING		
INFANT MORTALITY		
ANTENATAL CARE	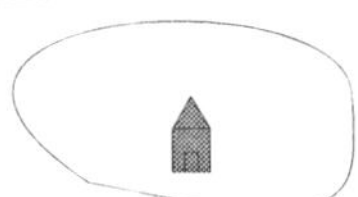	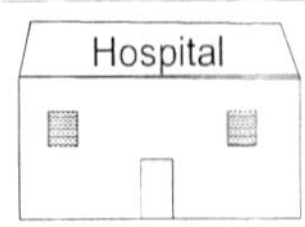
WOMEN'S ROLE	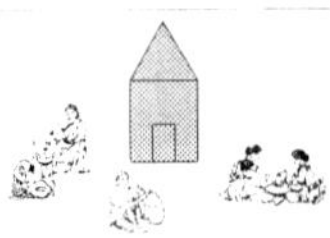	
POLITICAL POWER AND DECISION MAKING	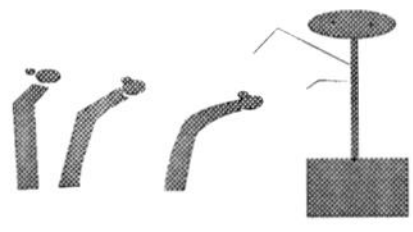	

Inferences

Inaugurated by Shri M.Inference that could be drawn from the three cases are:

- Participatory Rural Appraisal methods could be combined with other methods like Logical Framework Analysis.
- Combinations of two or three methods are likely to produce the required data for formulation of projects.
- The NGO completely relies on PRA as they are most relevant, reliable method.
- Method used for appraisal could also be used for monitoring and evaluation. For example, mapping used for appraisal was also used for evaluation; similarly the trend change. Participatory methods provide vast scope for innovation and improvisation.
- The methods empower the participants especially the women to discuss the issues/problems that affect not only their family but also the village.
- 90 per cent of the women are in a position to read and write. Continuous interaction through group discussion, guided dialogue, visual presentation, shared presentation have largely helped in enhancing their skills in reading and writing.
- The women could generate common fund through income generation programme like dairying.
- Income of the women has increased over a period of time.
- The role of moneylender has been minimised.
- Women have taken up the responsibility of running the ration shop.
- Some women groups have fought against corruption at the grassroots.
- Caste difference has narrowed down among the members of the group.
- Greater awareness on education, health and sanitation could be noticed.

TAMIL NADU WOMEN'S COLLECTIVE

The Women's Collective was established in June 1994. But the beginning can be traced back to the 80s. Many of the members of Women's Collective earlier (80s) were working as animators/ community organizers in small community-based NGOs in several parts of South India. All the members have, therefore, prior experience of community and organization work. Though the members work in different organizations, they know each other quite well and that led to form an informal support group known as "Voluntary Organizations Women's Staff Association". The purpose of the group is to find a space for sharing and helping each other regarding personal and organizational problems. Later, many of these women left NGOs where they worked and started their own. They identified their organization as women's organization; yet they lack proper gender perspective. They were committed and strongly inspired to work for a transformed society. Yet they were constrained in terms of (i) lack of solid analysis and understanding of community situation, particularly the situation of women. (ii) want of more focussed direction of what they want to achieve (iii) paucity of technical, material and financial resources.

Funding was a major problem. This has brought many of these women NGOs together mainly to have access to funds. But it did not end there. It attained more strategic stage, when each one of them slowly awakened to the idea of working together as a collective. Such a collective, it was felt, would transform the limitations into strengths. It would help them in mutual learning and support. The groups did not have previous experience of how a collective works. But they were prepared to experiment, evolve and develop a collective. That is how Tamil Nadu Women's Collective was born.

As a newly formed group, the initial task was to formulate basic policies known as the code of conduct. The code was subjected to lengthy discussions incorporating everyone's suggestions and ensuring that it was acceptable to every one.

Side by side, they started equipping themselves by undergoing training in participatory approaches and gender perspective, which helped them to go in for a thorough situation analysis of the villages where they worked. The first proposal using participatory approach submitted to 'Christian Aid' was approved in December 1994 six months after the birth of Tamil Nadu Women's Collective. The present core group of Tamil Nadu Women's Collective (WC) consists of 31 member organizations which extend to 11 districts covering 450 villages across Tamil Nadu. These organizations have been responsible for the formation of 526 Self-Help Groups and 411 Women Sangams with a membership of 8000.

The objectives of Tamil Nadu Women's Collective are to:

— facilitate the members to conceptualize the issues and problems using action-reflection-action process.
— help the members in establishing linkages with community, state, and regional level networking.
— help the women to resolve conflicts and problems arising from caste, class and gender oppression.
— help the members to practice the art of planning with community women using participatory approaches.
— facilitate the formation of women, youth, children's self-help group through education, awareness raising and skill training.
— assist the federation of women sangams to strengthen community-based women's groups.
— enhance the role of animation cultural team through training and exposure programme.
— enable women's sangams to take the initiative to address gender based and sectoral issues.
— undertake participatory research and documentation of gender and sectoral issues to be addressed by women.
— ensure food security at the family and village level.

Four major functions to realize the objective are:

- campaigns, demonstrations, solidarities and networking;
- formation of sangams and savings/credit schemes in the village;
- holding women's conferences and exposure visits; and

- land development and food security programme.

The regular work revolves around facilitation of everyone's participation. The Collective strongly believes that everyone has an equal voice and equal right to decide, plan and act in all matters relevant to them as individuals and members of the community.

The daliths, tribal and other backward castes constitute primary focus of work in the villages.

To enable the members to undertake community work, the Women's Collective organized a number of training workshops, for both staff and sangam members. The focus of the trainings is philosophy of participatory work and the use of Participatory Rural Appraisal (PRA) techniques. The training helps them to gather information on :

— village histories
— natural resource and environmental issues
— demographic details
— seasonal patterns and issues
— land use patterns
— identification of gender issues and other concerns of women
— identification of critical and urgent issues affecting women and their communities.

All members of staff have used PRA in their working area, which resulted in pinpointing issues affecting women and requiring attention. The use of PRA proved effective in :

- building people's confidence about themselves;
- building trust on the members of the WC as outsiders;
- laying the foundation for friendship and solidarity;
- encouraging villages to have link with each other; and
- communicating mutually their needs and problems.

By applying the PRA principles and using its techniques, women in the community were able to plan programmes on their own, which include :

- awareness raising and training
- exposure visits

- sangam management
- assessment of village needs.

The WC staff facilitate the process where the communities themselves can use the PRA techniques. Some of the techniques that were used in the communities include:

- — village mapping
- — linkage mapping
- — time lines
- — chappathi diagram
- — seasonal clocks
- — 24 hour division of labour clock
- — problem solving maps
- — trend change over time chart
- — access and control matrix
- — interest outflow map

Inferences

Tamil Nadu Women's Collective is an example for institutionalizing the concept of PRA for it has practiced basic philosophy at the institution itself—the philosophy of openness, transparency, non-domination, near elimination of hierarchy, sharing etc.

- empowering the community by making them do PRA exercise independently.
- programme planning by people themselves using participatory methods.

Conclusion

The analysis in this chapter is mainly based on NGOs experiences in practicing PRA methods. NGOs under study resort to PRA as it enhances and reinforces participation of the people at different stages of the project. PRA is found to be an effective method and approach at different stages of the project cycle. PRA has been viewed as less costly, less time-consuming and user-friendly in gathering information, in formulating projects and in executing and evaluating the projects. Hence, NGOs apply PRA at all phases of the project cycle and development intervention.

PRA has a menu of several methods. The NGOs under study

have picked up and used a few methods in their development work. The uses of methods among NGOs do vary. For instance, the methods frequently used by OUTREACH are venn diagram, matrix scoring and 100 seeds scoring. SPEECH's popular methods are mapping of different types, seasonal calendar, trend change, and venn diagram. ASSEFA uses social mapping, wealth ranking and venn diagram. All the NGOs improvise the methods depending on local conditions and situation.

The experience of NGOs clearly indicate that participatory methods empower the people particularly the women and marginalised in areas of gathering and analyzing information, in decision making, in implementing and evaluating projects.

Participatory methods, as could be seen from the experience of NGOs, have released the poor from the shackles of dependency and led them to independence as the methods have helped in designing and establishing micro-enterprises. The methods have thus resulted in economic empowerment. The methods have made the participants to realize the importance of self-development.

NOTES AND REFERENCES

1. Andrea Cornwall and Rachel Jewkes, "What is Participatory Research", *Social Sciences Medicine,* Vol. 41, NO. 12, 1995, p. 1668.
2. Quoted in Amitava Mukherjee, *Participatory Rural Appraisal Methods* and *Application in Rural Planning,* Vikas Publishing House Pvt. Ltd., New Delhi, 1995, p. 65.
3. Biggs, S. "Resource-Poor Farmers' Participation in Research: A Synthesis of Experience from Nine National Agricultural Research Systems", *OFCUE Comparative Study Paper 3,* International Service for National Agricultural Research, The Hague, 1989.
4. Robert Chambers, *Rural Appraisal: Rapid, Relaxed and Participatory,* Institute for Development Studies, Discussion Paper No. 311, Brighton, 1992.
5. Scott R.A., and Shore A.B., *Why Sociology does not Apply: A Study of Sociology in Public Planning*, Elsvier, New York, 1979.
6. Amithava Mukherjee, *op. cit.*, 67.
7. Robert Chambers, *Whose Reality Counts? Putting the First Last,* Intermediate Technology Publications, 1997, p. 199
8. Robert Leurs, "Current Challenges Facing Participatory Rural Appraisal", *Public Administration and Development,* Vol. 16, No. 1996, pp. 61-62.

5

EFFICACY OF PARTICIPATORY METHODS

Introduction

Participatory Rural Appraisal (PRA) as a methodology has primarily been evolved to appraise the rural problems and conditions. The evolution of the concept authenticates this. PRA was originally defined as "an approach and method for learning about rural life and conditions from, with and by rural people"[1]. In course of time, the profounders of this concept felt that it is not mere learning about rural conditions. As a development tool, it cannot just stop with committing the people in analyzing their problems, conditions and situations. It must go beyond that and "extend into analysis, planning and action." The description of PRA therefore has to be refashioned to cover these aspects. PRA in the changed context has been defined as "a family of approaches and methods to enable rural people to share, enhance and analyze their knowledge of life and conditions, to plan and to act"[2].

The ails of Conventional Approach

The change in the concept of PRA from mere appraisal method to all inclusive action-centered technique was mainly because of its rapid spread, especially among NGOs in Third world countries and also its applications in different fields of development activities[3].

Further, the disillusionment and disappointment of social scientists with the conventional methods of data gathering, planning and ultimately the development have positively backed the refashioning, the concept of PRA. The top-down nature of the conventional approach, the social scientists have realized, has done enough damage to the pursuit of rural development and failed to answer the phenomena of poverty, unemployment, inequality and ecological degradation. It is being increasingly recognized that the conventional methods are neither realistic nor sustainable in the long-run[4]. This is to a larger extent true because our development thought and action has largely served to legitimate. (a) centralised authoritarian governments as the instruments for achieving optimal allocation of development resources; (b) control-oriented administrative structures notorious for their inability to innovate and to deal with the dynamic variety of complex and changing societies; and (c) projectized resource allocation schemes that provide fragmented and sporadic bursts of development action to the neglect of system interdependence and sustainability, and draw attention away form the application of experience to improve the performance of development programme"[5].

Power to the People

The solutions to many of the ills in the traditional authoritarian development theories, it is strongly believed, remain in empowerment-oriented people-centered participatory approach which calls for a "return of decision making power to the people, acknowledging the legitimacy of their individual preferences and genius of their ability to mobilize available resources to address their self-defined needs"[6]. The concern here is less with decision-making techniques that will yield the one optimal answer, but more with the processes by which many people interact to arrive at mutually acceptable solutions consistent with their varied and often conflicting interests[7]. This entails a redefinition of the role of bureaucrats from that of implementor or of predetermined procedures to that of problem solver responding in creative ways to facilitate both planning and action by the people themselves.

The Spread of PRA among NGOs

NGOs, who have been working with the people, have quickly responded to the participatory approach and started practicing "seemingly simple participatory methodologies involving disciplined observation, accurate recording and common sense analysis"[8]. To start with, they practice the methods of PRA at the grassroots which may "provide basic information on the feasibility of beginning a project"[9]. Later, realising the efficacy of the methods experimented the methods in the area of analysis of situation, planning and implementation of projects. The results of experiments were found to be encouraging, rewarding and impressive which eventuate in the quick spread of PRA in all corners of the globe". The reasons for quick spread are many. Among them some stand out. First, the sort of open institutional culture in which PRA has evolved and thrived has become common among NGOs. Openness has inspired innovation. Secondly, the practical applications have contributed to the momentum. Third, PRA proved enjoyable and generated rapport. Fourth, the information and insights which flow from it have often been diverse, detailed, complex, accurate, interesting and useful and shared in a short time. Fifth, PRA has again and again proved both useful and powerful[10] and offer beneficial strategy for participatory development.

The widespread use of PRA in diverse contexts signals the extent of its appeal. The practitioners of PRA often claim that use of PRA in some settings have, *inter alia*, helped to (i) empower marginalised communities and groups, by encouraging them to analyse local conditions, giving them confidence to assert their priorities, to present proposals, to make demands and to take action; (ii) pursue community-based processes for development including appraisal, planning, implementation, monitoring and evaluation; (iii) seek and enable the expression and integration of local diversity in otherwise standard programmes; and (iv) identify local priorities for research and initiate participatory research with scientists becoming more receptive to local knowledge and recognising that farmers are able to design, conduct and evaluate their own experiments.[11]

Dangers, Challenges and Criticisms

The rapid spread and application of PRA has exposed them to certain dangers; have made them face certain challenges; and has invited sharp criticisms too.

The dangers,[12] as vividly exposed by Robert Chambers, may be stated. The first is instant fashion. PRA is vulnerable to over rapid promotion and adoption, followed by misuse, and by sticking on labels without substance.[13] The second danger is rushing. The use of the word 'rapid' (Rapid Rural Appraisal) has become a liability. It has been used to legitimize brash and biased rural development tourism. Hurried rural visits, insensitive to social context, and lack of commitment compound errors and can mean that the poorest are once again, neither seen, listened to, nor learnt from.[14] A third danger is formalism. With any innovation, there is an urge to standardize and codify often in the name of quality. Manuals are called for and composed. They can really be beneficial as compiliation of ideas and experience, as handbooks that widen the choice of methods and applications, and as source of tips and applications. But manuals do inhibit and intimidate[15] and restrict innovations. A fourth danger is routinization and ruts. PRA is being scaled up. There has been a great demand for PRA from several corners. This results in repetition of few exercises by practitioners and trainers. Repetition thus leads to regular habits[16]. These dangers put together are likely to undermine the quality of PRA.

The challenges are many. The first is related to hierarchy. PRA presupposes that the members of team practising PRA should forget hierarchy in the field and work on equal footing. In reality, it does not always happen and there is certain degree of hierarchy among them. The second challenge is with regard to raised expectations. PRA created dependency culture. Conducting an organised PRA exercise involving a group of outsiders demonstrate a visible commitment on the part of the project to a particular community[17] which in turn raises the expectations of people[18]. Thirdly, the participation in the initial period is poor both in terms of quality and quantity. Again, once participation is secured, involvement of the people in the process of appraisal is usually neither continuous nor predictable. Commitment and interest wanes overtime. Participants can experience task exhaustion and the composition of appraisal group fluctuates overtime.[19] Fourthly the dominance of males' view still pervades the exercise of participatory appraisal. Women are yet to fully participate in the appraisal. Organised group PRA exercises have not provided appropriate contexts for the articulation of women's perspectives. This exclusion of women is 'so normal' and naturalised' that is rarely noticed nor questioned. In fact, the presence of women causes remark while their absence go unnoticed[20]. The practitioners of PRA do face all these challenges in the field.

The criticisms are: (i) they are biased, impressionistic and unreliable[21]. They constitute an enquiry that is undisciplined and sloppy. They are said to involve subjective observations and respond fast to the selected members of the community. (ii) terms like informal and qualitative are used to imply poorer quality or second-rate work. (iii) rigor and accuracy are assumed and therefore to be in contradiction with participatory methods.[22] (iv) PRA is prone to poor and gender sensitive. Commitment to work with poor and powerless is part of the process. However, the poorest, marginalised, inaccessible, unorganized and fragmented sometimes are easily left out. There are plausible reasons for this. The attractive methods of PRA can easily be dislocated from methodological framework and put to serve within and beyond community[23]. PRA is time intensive process. Therefore those with least amount of time (the poorest, particularly the women) are mostly left out of public forums where much of PRA activity (analysis and decision making) is carried out[24]. Participatory appraisal is more easily facilitated through the medium of domestic stakeholders who often must be able to mobilize resources and articulate concerns. The poorest and marginalised are rarely represented among them.

The dangers are imminent, the challenges do exist; the criticisms are fair and constructive.

Superiority of PRA

The dangers, challenges and criticisms revolve around problems and difficulties faced while applying the participatory appraisal methods in appraising a situation or condition, planning, execution, monitoring of a project/scheme. The criticisms arise especially when PRA is claimed as a methodology superior to conventional survey method. The propounders of the concept of the PRA cite shortcomings in 'survey' and 'rural development tourism' as the major reason for the evolution of participatory methods. Robert Chambers describe them as 'negative sources' of PRA. Rural development tourism cannot be termed as scientific system of data gathering as it is a totally imperfect application of the principles of observation and discussion. But survey questionnaire is a scientific and widely used technique for data collection. In fact, large scale questionnaire surveys, whether for one-off ad-hoc investigation or for longitudinal studies remain one of the most widespread and sustainable rural industries. However, it suffers from serious shortcomings. The shortcomings of the survey are: (i) it is costly; (ii) it is time-consuming; (iii) cross-checking of information is not possible; (iv) data are not shared with respondents; (v) power and initiation lie with the interviewer; (vi) it is rigid, pre-determined interaction process without scope for flexibility and (vii) the concern is less with the people (respondents) and more with what they provide (responses). The responses matter more, for they are raw material to be mined, packaged, transported and processed, the commensurable output to be collected, categorized, coded, counted and correlated.[25] PRA methods and techniques, it is claimed, do away with such shortcomings and fulfill the criteria of effective information supply. The criteria for effective information supply[26] are:

- the data and information should be wholistic and comprehensive.
- they must be cost effective.
- they must relate to target and goals.
- they must be operationally relevant.
- they must be timely

- the system must deliver information direct to the appropriate levels of management.

Do participatory rural appraisal methods fulfil these criteria? The answer to these questions will help us to prove whether participatory methods are affective or not. The answers may be found at two levels. (i) at the level of the field workers and practitioners of PRA; (ii) at the field itself.

Efficacy Criteria

We present an analysis of the views/opinion of the staff of the selected NGOs who have been practising PRA sufficiently for a longer period of time and at different stages of project cycle. The criteria considered for studying the efficacy of the PRA methods are:

- Reliability
- Timeliness
- Rapidity
- Cost effectiveness
- Simple to analyse
- Easy to practice
- Accuracy
- Appropriateness
- Participation
- Adequacy
- Sharing of information
- Creation of positive attitude and behaviour

These criteria to test the efficacy of PRA methods in appraisal have been largely developed based on; (i) extensive review of literature; (ii) discussion with the practitioners; and (iii) the experience of Gandhigram Rural Institute.

Each of the criteria may be briefly stated

Reliability refers to repetition of findings if the inquiry was replicated with the same subjects in the same or similar context.
Accuracy refers to the exactness of the data collected.
Timeliness refers to availability of data and information on time without any delay.
Appropriateness means relevance of the data or conformability to a specific situation or condition.
Rapidity means quickness with which data could be gathered.
Participation means not mere physical participation but active

participation of the people at different stages of information gathering and sharing.

Cost effectiveness means collection of primary data at a lesser cost when compared to other methods of data gathering.

Adequacy means enough or sufficient data for the purpose for which they were collected.

Simple to analyze refers to analysis of data without problems.

Positive attitude and behaviour means generation of certain traits like openness, transparency, non-domination, humility and listening, while practicing and after completing PRA exercise.

Sharing refers to sharing of data or information with the participants (insiders and outsiders).

Easy to practice refers to practicing the methods in a given situation with ease.

Efficacy as perceived by NGOs

The method adopted to assess the efficacy of PRA methods among the NGOs may be stated.

- Each criterion was explained in a gathering of NGO staff who have been practicing PRA methods in the field.
- Each criterion was assigned a score of 10 without any discrimination among the criteria, as all of them are equally important.
- Every participant in the gathering was asked to assign the score for each criterion out of the maximum score of 10. Thus, every participant was asked to assign score for all the 12 criteria.
- The scores given by all the participants for each criterion was totalled and averaged.

The results are presented in Table 5.1

SPEECH (Thiruchuli, TN) has found PRA methods more effective. PRA methods are absolutely effective in terms of 'reliability', 'accuracy' 'appropriateness', 'adequacy' 'simple to analyze' 'sharing and 'easy to practice'. It has also expressed its strong faith in other criteria. OUTREACH (Bangalore) too expressed its faith in the efficacy of participatory methods. However, it has not fully agreed with efficacy of PRA in terms of 'rapidity'.

Table 5.1: Efficacy of PRA as reported by NGOs

Variables	*NGOs*					*Total Score*	*Average Score*
	OUTREACH	*SPEECH*	*ASSEFA*	*SEED*	*RENAISSANCE*		
Reliability	9	10	9	8	6	42	8.4
Accuracy	9	10	7	9	5	40	8.0
Timeliness	10	9	8	8	8	43	8.6
Appropriateness	9	10	9	9	7	44	8.8
Rapidity	7	9	8	6	6	36	7.2
Participatory	9	9	9	8	9	44	8.8
Cost-effective	8	9	8	8	9	42	8.4
Adequacy	9	10	8	9	7	43	8.6
Simple to analyse	9	10	9	9	9	46	9.2
Positive attitude & behaviour	9	9	8	9	6	41	8.2
Sharing	10	10	9	9	7	45	9.0
Easy to practice	10	10	7	9	8	44	8.8
Total Score	108	115	99	101	87	510	—
Average	90	95.8	82.5	84.2	72.5	85	—

SEED and ASSEFA have more or less assigned equal score. They too considered PRA methods to be effective. ASSEFFA, however, is not fully convinced of the efficacy of PRA with reference to criteria like "accuracy" and 'easy to practice'. SEED has its reservation on the efficacy of PRA with regard to "rapidity".

RENAISSANCE has given the least score. They believe in efficacy of PRA in terms of "participation" 'cost effectiveness' and 'simple to analyze'. It has its reservation on criteria like 'accuracy', 'reliability', 'rapidity', 'creation of positive attitude and behaviour'.

The overall score indicates that participatory methods are quite effective with reference to all the criteria barring 'rapidity'.

The reasons for the difference in the scoring by different agencies are obvious. The agencies like OUTREACH, SPEECH have been constantly practicing PRA over a decade at different stages of project cycle. The staffs are well trained and are exposed to a wide variety of situations. SEED and ASSEFA too practice PRA for quite a long period of time. But they apply PRA in limited and restricted fields of activities. They are yet to have a full grip over the methods. RENAISSANCE applies PRA in just appraisal alone. Its exposure to PRA is also limited. Hence, it is yet to be fully convinced of the efficacy PRA.

Efficacy of PRA as experimented

The discussion so for centered on the efficacy of PRA as perceived by the NGOs with reference to select criteria. The NGOs especially the field workers who have been practicing PRA for quite a long period of time have reported that PRA methods have been quite effective. Based on NGOs' experience alone, we may not be able to come to a definite conclusion about the efficacy of PRA methods. Further, there needs to be more investigation into the accuracy of information gained through PRA and RRA exercises, especially in comparison with other methods of data collection.[27]

We now make an attempt to analyse the efficacy of participatory methods in terms of 'cost-effectiveness; 'rapidity', 'accuracy', and 'participation' based on the experiments conducted in the field. First, we choose two criteria as RRA and later PRA mainly evolved as cost effective rapid methods. We have added accuracy as a third criterion mainly to prove that quantitative data collected through PRA are as accurate as they are collected through survey method. We also added 'participation' in our analysis as participation in PRA has been questioned.

We have, for the purpose of the experiments, chosen three different villages. Each village is distinctly different from the other in terms of size, caste, structure of settlement, occupation etc. For instance, one village (Chellakuttiur) is predominantly of nomads with joint family system very much intact. The other two villages are smaller in size. Yet, a lot of difference could be seen in terms of caste composition, occupation and structure of settlement etc. The purpose of choosing three different new villages for the experiment is to see that size and homogeneous nature of the population do not have much influence on the participation. For example, our experience of organizing PRA in more than 100 villages of different size and composition have conclusively proved that the size of the village in terms of number of households is a major determinant in enlisting people's participation and thereby quickly completing the appraisal.[28] Similarly, a homogeneous village in terms of caste composition provides a very enabling environment to complete the appraisal as rapid as possible. Conducting experiments on the effectiveness of PRA vis-à-vis survey in such small, homogeneous village is likely to bring biased results. Hence, we have chosen a combination of three villages which are different in nature. We also made it a point to see that the villages are new, remote and PRA exercises were not conducted earlier. These precautions were mainly taken to see that the results obtained out of the exercise are not biased.

The methods chosen to compare the data are social mapping under PRA and conventional survey methods. The reason for selecting the participatory mapping from among the menu of methods is that participatory mapping has now been increasingly used as an alternative for collecting the demographic and other quantitative data.[29] The type of data collected under the two methods are:

Name of the head of the household
Religion
Caste
Marital status
Age of the family members
Literacy status of the family members
Occupation
Type of house
Extent of land owned

- Possession of livestock such as bullock, cows, buffalo, goat, sheep and others.

The procedure followed for collecting the data under two methods may be stated. A team of two investigators, after proper orientation, has been deployed in the field to gather data under survey method. They have been asked to note the time taken to collect the data from each household. The same team was given the responsibility of analyzing the data and reporting the analysis. They were also asked to note the time taken to complete the data analysis and reporting. Thus, the total time taken to complete the task under survey was arrived at.

For organizing PRA, a team of three members trained in PRA were deployed in the field. They were entrusted with the responsibility of gathering information using participatory mapping. The data as collected through survey was gathered through mapping too. The team was asked to note the time taken to complete the process of data gathering, analysis, reporting and sharing.

We will now present the time involved for collecting the data under the two methods.

Rapidity (Time involved): The factors to be considered for the time taken to complete the process of data collection, analysis and reporting are: [30]

- training time
- time taken for conducting PRA/Survey
- time required for the kind and size of data to be collected
- time taken for cross checking of data

- time required for data management
- time taken for dissemination of data

The steps involved in conducting PRA method which influence time and duration are: (i) kind of data required; (ii) rapport building; (iii) techniques whether 'visual' or 'verbal' or 'a combination'; (iv) probing on various issues; (v) copying of 'visual' output on paper; (vi) taking of notes; and (vii) cross checking of data.[31]

We have not considered the training time as we have deployed already trained personnel for conducting PRA and also for survey. All the other factors have been carefully considered and incorporated for analyzing the time to collect a set of quantitative data under PRA and survey.

The time involved to collect data under the two methods clearly shows that survey has taken 299 per cent of time than PRA for all the three villages put together. The time taken to complete a survey per household is 14.40 minutes while for the PRA, it is 4.81 minutes for all the villages put together.

Village-wise analysis of the time taken under two methods, presents more or less similar trend in the sense that survey consumed around 300 per cent of the time than PRA. However, there are variations in the time taken across the villages. For instance, the time taken to complete the survey per household for Chellakuttiur is 15.50 minutes and for PRA 5.04 minutes; whereas for other two villages the survey time per household is much less and more or less same and so with PRA too. The survey time, for instance, for Pannaikulam is 12.98 minutes and for Karuthakkapatti 12.53 minutes. The time taken for PRA at Pannaikulam is 4.45 minutes and at Karuthakkapatti 4.50 minutes.

The variation in the time taken in the first village and the other two villages may be explained. The first village is a comparatively bigger village. The bigger village is likely to pose problems like building-up rapport, enlisting participation of different groups of people, finding the right key informants, bringing several kinds of leaders around, etc. All these factors have influence on duration of PRA. Hence, it is understandable that PRA in a comparatively bigger village takes much time. But the size of the village does not have any influence on the time taken to complete the survey. Yet,

Table 5.2: Time taken to gather data: A comparison between Survey and PRA Methods

(in minutes)

Details	*Chellakuttiur*		*Pannaikulam*		*Karuthakkapatti*		*Total*	
	Survey	*PRA*	*Survey*	*PRA*	*Survey*	*PRA*	*Survey*	*PRA*
Rapport building	290 2.28	180 1.42	106 2.30	40 0.87	94 2.35	35 0.88	490 2.30	255 1.20
Data collection	1029 8.10	240 1.89	241 5.24	90 1.96	199 4.98	75 1.88	1469 6.90	405 1.90
Data analysis	490 3.86	120 0.94	190 4.13	40 0.87	152 3.8	30 0.75	832 3.91	190 0.89
Reporting	160 1.26	60 0.47	60 1.30	20 0.44	56 1.4	25 0.63	276 1.30	105 0.49
Sharing	— —	40 0.31	— —	15 0.33	— —	15 0.38	— —	70 0.33
Total	1969	640	597	205	501	180	3067	1025
Time spent per unit	15.50	5.04	12.98	4.46	12.53	4.5	14.40	4.81

survey is found to have taken much time in the big village. Why? That is mainly because of the big size of family in Chellakuttiur which is again mainly due to wide prevalence of joint family system in the village. For instance, there were 52 members in a household at this village. This is further supported by the fact that the average family size in this village is 7.5 whereas in the other villages, it is 4.6 each.

The time taken per unit (per household) under different components like rapport-building, data collection, data analysis, reporting and sharing shows that data collection takes much time both under survey and PRA followed by data analysis, rapport building and reporting. We have not shared the data collected out of survey with the respondents.

In final analysis, it can be said that PRA takes much less time than the survey method whether in a big or small village. The findings of an earlier study corroborate this.[32] The time taken by PRA in different settings does differ. Big village takes much time to complete PRA exercise (mapping) whereas a smaller village takes less time.

While comparing the time taken to collect data under two methods, we have included the researchers' timing only. We have not included the time spent by the participants from the community. Michael Richards *et al.* argue that time is one of the main factors in an analysis of cost effectiveness. If community time is not included and/or when no compensation is offered, PRA is cost-effective. They report that comparing researchers/facilitators and community time makes PRA absorb approximately 240 per cent more of every one's time than the survey.[33] We have not considered community time for obvious reasons. Firstly, we organized the PRA at their convenience. Secondly, we assured of ourselves that they do not have work during the period and have not lost wages. Thirdly, it is a kind of mutual learning. Fourthly, such an exercise brings solutions to some long-standing problems.[34] This naturally leads to assessing the benefits, which the participants derive out of such exercise. In other words, while assessing the cost, the benefits should also be taken into account. And therefore, we need to work out the net cost and the net benefit of the process. Finally, our long years of experience in organizing PRA, have shown the participants never complained of losing 'something' while attending PRA. On the other hand, they have quite often reported to gain 'awareness' and 'knowledge' about their life and conditions.

Further, there are certain other advantages of participatory mapping. They are as follows: (i) data are generated with completion of such mapping and are readily available at one location as compared to conventional census listing of households where data are collected from door to door and are available only after its formal processing and publication; (ii) both the researchers and participants from the community can have a map of the village which otherwise is very difficult to get; (iii) the maps so drawn by the villagers can further be used for PRA exercises like wealth ranking and some times for future visioning. The exercise gives them the confidence that they can also do it; (iv) the information is their's. They express it, own it and share it; (v) they often enjoy the creativity of what they are doing and what they see and learn through their presentation and analysis. Thus, the multiple benefits of PRA exercise like mapping the community is likely to outweigh the 'cost', if at all any, borne by the community.

Cost Effectiveness

Robert Chambers described survey as a long and dirty method, which means data collection in a large questionnaire survey, can be costly and time consuming. The practitioners of PRA claim that it is cost effective (see for example, the score given by NGOs in the same chapter for cost-effectiveness). How do we arrive at cost? It is not easy to arrive at the relative cost for two methods through simple comparison, as there are many factors, which influence both actual cost and imputed cost of undertaking conventional survey or PRA type of studies. Such factors are:

- Planning and design for data collection
- Training in PRA or conventional survey
- Size of the area to be covered
- Mode of transport
- Materials used in data collection
- Number of investigators/facilitators

- Duration of enquiry, methods, analysis and results
- System for dissemination of results.

For calculating the cost, we have considered three broad components namely, manpower charges, printing and stationery and documentation. Calculation of time for different stages of data collection, analysis and compilation, largely helps in computing the manpower charges involved in the process. The cost on other two components can be easily worked out. The village-wise cost of data gathering under two methods is given in Table 5.3.

The total cost of conducting survey of 213 households works out to Rs.2, 340/- whereas the cost of collecting the same data under PRA is Rs.964/- Thus, the cost of conducting survey for a set of census data was 243 per cent more than the cost incurred on PRA for collecting the same data.

Village-wise analysis of the cost shows that the cost of conducting survey and PRA was higher in Chellakuttiur than the other two villages. The reason is quite obvious. The time taken to complete the survey and PRA was high in the village and therefore the highest cost. Again, village-wise comparison of cost data (total as well as unit cost) shows that the cost of organizing survey was 262 per cent more than PRA in Chellakuttiur, 216 per cent in Pannaikulam and 214 per cent in Karuthakkapatti. The results of the study have clearly shown that organizing PRA for collecting census data is less costly.

Accuracy

Participatory methods have been criticized mostly on the ground that it lacks rigor. The purpose of rigor is trust-worthiness.[35] Trustworthiness of data largely depends on the accuracy of data. The survey data are considered to be trustworthy, accurate and are amenable to the very powerful and rigorous techniques of modern statistical analysis".[36] Whereas PRA data, it is often said, are not amenable to the rigor and accuracy. This criticism to a certain extent has been answered by applying the conventional tests which involves comparison PRA approach with others (See review of literature). Robert Chambers has identified four main

Table 5.3: Cost of data gathering: A comparison between Survey and PRA methods

(in Rupees)

Cost details	*Chellakuttiur*		*Pannaikulam*		*Karuthakkapatti*		*Total*	
	Survey	*PRA*	*Survey*	*PRA*	*Survey*	*PRA*	*Survey*	*PRA*
Manpower charges	900	500	275	150	240	150	1415	800
Printing & Stationery	375	48	65	20	65	19	505	87
Documentation	300	52	60	15	60	10	420	77
Total	1575	600	400	185	365	179	2340	964
Unit Cost	12.40	4.72	8.70	4.02	9.13	4.47	10.99	4.53

areas where RRA and PRA have generated numerical data which can be compared with those from conventional sources. They are: farm and household surveys, wealth and well-being ranking; participatory mapping and village census, and rainfall.[37]

Field studies have proved that the data, especially quantitative data, collected through PRA are as accurate as data collected through conventional methods.[38] Such field studies are limited. Substantial and convincing evidence need to be produced through more rigorous field studies to prove the efficacy of PRA over conventional method. We made further attempt to test the accuracy of data collected through PRA in comparison with survey method. Enough care has been taken to see that the data collected through PRA have been triangulated. The data collected through both the methods are presented in Table 5.4.

Regarding the number of households and type of houses, survey and PRA output provide the same results and there are absolutely no differences. There is an insignificant variation in population especially in the case of big village-the variation being three. The survey results were found to be accurate on further verification of the data by comparing PRA and survey data house by house. In the case of small villages, PRA data perfectly matched with the survey data. Even such minor variations could be avoided through careful triangulation.

The data on landless were similar in both the cases. But there are variations when the data on extent of land owned were examined. The variation is quite negligible. Here again, the survey data was found to be more accurate than PRA data. This is also understandable as the data provided by participants of PRA are 'near accurate' data.

Regarding the data on cattle population, there is very slight difference between two sets of data. Here, PRA data proved to be more accurate as some of the respondents during survey have not counted 'calves' under cattle population whereas under PRA, all the animals including 'calves' have been considered.

Table 5.4 : Comparison of data collected through Social Mapping and Survey Method

Details of Data	*Chellakuttiur*			*Pannaikulam*			*Karuthakkapatti*		
	Survey	*PRA*	*Variance*	*Survey*	*PRA*	*Variance*	*Survey*	*PRA*	*Variance*
1	2	3	4	5	6	7	8	9	10
No. of households	127	127		46	46	—	40	40	—
Type of house									
Thatched	36	36	—	15	15	—	14	14	—
Tiled	72	72	—	25	25	—	19	19	—
Concrete	12	12	—	6	6	—	7	7	—
Colony	7	7	—	—	—	—	—	—	—
Population	954	957	3	215	215	—	184	184	—
Male	471	472	1	110	110	—	92	92	—
Female	483	485	2	105	105	—	92	92	—
Land details									
Landless	71	71	—	21	21	—	23	23	—
< 2 acres	33	36	3	7	8	1	8	8	—
2-4 acres	14	13	1	11	10	1	4	5	1
4-6 acres	6	4	2	5	5	—	4	4	—
> 6 acres	3	3	-	2	2	—	1	—	1
Cattle population	238	241	3	236	235	1	78	78	—
Cow/Buffalo	105	110	5	55	54	1	33	33	—
Goat/Sheep	133	131	2	181	181	—	45	45	—

Contd......

Table 5.4 *(Contd.)*

1	2	3	4	5	6	7	8	9	10
Age (years)									
0-5	157	156	1	19	19	—	9	9	—
5-14	190	191	1	41	40	1	18	17	1
14-35	374	377	3	90	91	1	79	80	1
35-59	181	181	—	45	47	2	62	60	2
59 +	52	52	—	20	18	2	16	18	2
Literacy									
Illiterates	562	560	2	92	92	—	47	47	—
std. 1-5	207	211	4	27	27	—	39	39	—
std. 6-8	107	106	1	39	39	—	40	40	—
std. 9-10	45	47	2	32	32	—	33	33	—
std. 11-12	24	23	1	22	22	—	23	23	—
std. > 12	9	10	1	3	3	—	2	2	—

Age-wise classification of the total population under the two methods also revealed a slight variation in the data. The data collected through survey was found to be more accurate.

Regarding literacy, the data collected under the two methods were accurate and exactly tallied even in the case of small villages. There is slight variation between the two sets of data in Chellakuttiur and the survey data was found to be more accurate on further crosschecking.

It can be concluded that data collected through PRA (quantitative) could be as accurate as data collected through survey.

Participation

'Participation' and 'participatory' are the frequently used words in development. They have variety of meanings. Participation is the voluntary involvement of people in self-determined change. It is involvement in people's development of themselves, their lives and their environment.[39] It is a process, which allows the people to exert influence over their lives and the conditions under which they live.

Participation of people in projects has been emphasized for several reasons. Prominent among them are: (i) participation may result in better decisions. People have information which the outsiders lack. Further, participation permits different views to be aired and in this way, the danger of group thinking is reduced; (ii) people may be more likely to implement decisions they helped make themselves than decisions imposed on them from above. Not only do they know better what is expected of them, but helping them to make a decision commits them to it; (iii) motivation is frequently enhanced, psychology has shown, by the setting up of goals during the participatory decision making process; (iv) participation may improve communication and cooperation. People may cooperate with each other and with the project staff. Further, by disseminating the experience in people's problem solving, participation may facilitate organizational learning. In doing so, participation contributes to dynamic efficiency;[40] (v) on a personal level, people may learn new skills through participation; leadership potential may be readily identified and developed.[41]

There are two major alternative uses of participation. They are (i) participation as an 'end' in itself; and (ii) participation as a 'means' to development. As an end, participation entails empowerment i.e., everybody's right to have a say in decisions concerning their own lives. As a means, it leads to efficiency in project management i.e., participation is a tool to implement development policies. Participation as an 'end' represents 'transformational participation' and as means represent 'instrumental participation'. Transformational participation is when participation is viewed as an objective in and of itself, and as a means of achieving some higher objectives such as self-help and/ or sustainability. Instrumental participation is when participation is viewed as a way of achieving certain specific targets-the local people in the outsiders project.[42]

Participatory Rural Appraisal aims at empowerment oriented or transformational participation. This would mean active participation of people at different stages of the project viz., appraisal, analysis, planning, execution, monitoring and evaluation. Empowerment starts at the appraisal stage itself. Participants are empowered to gather data on themselves using mostly visual media. The data, thus, gathered are analyzed and shared among themselves and also with the outsiders. The data analyzed by them mostly help them to find out their problems and solutions thereat and thus enable them to take decision on further course of action. Participation of people at the appraisal stage is, therefore, crucial and critical. PRA stipulates participation of different categories of people-old and young, men and women, literates and illiterates and so on. More specifically, outsiders should see that views/opinion/ voice of the marginalised, oppressed, downtrodden, and the poor are properly and fully taken into account. But then this may not invariably happen? The reasons are many. Some are internal-related to the participants. Some are external in nature-related to outsiders (See chapter 3). We now present the participation of people in a participatory mapping.

The observation on the participation of people may be explained. The team, to start with, identified the key informants and invited them to participate in participatory mapping. A member of the team was given the task of observing the time at which the exercise was started and the time at which it was closed. He was also asked to note down the number of participants sex-wise, (at different time intervals) in terms of how many were at the beginning, how many joined at different time intervals and how many left. The data gathered through such observation is presented village-wise (see Table 5.5).

It could be observed from the table that there were just eight participants at the start of the exercise. It slowly increased and went up to 29 at the close 2 1/4 hours. And it came down to 18 at the close of the exercise. It may noted that at the start of the exercise, the men were more in number whereas at the close of the exercise, women were more in number. Though we follow the rule of 'closing the group', we could not strictly adhere to such rule as participants are likely to join the middle of the exercise and some are likely to leave. We cannot force them to remain throughout the exercise. Further, the village being comparatively bigger with a high population, it took much time for us to collect the data. As the time passed, the interest of some participants waned and hence they left. However, it should be noted that a majority of the participants stayed back till the exercise was over.

The other two villages are comparatively smaller. It took less time to complete the exercise. The team could sustain the interests of the participants. Only very few participants left in the middle of the exercises. There were 12 participants at the start of the exercise (in Pannaikulam) and went up to 20 at 2 1/4 hours and came down to 17 at the close of the exercise. Presence of women was more or less equal to men.

In the case of Karuthakkapatti, there were 13 at the beginning and it rose to 20 at the close of the exercise. Barring a lonely man and a woman who left at the middle of the exercise, all the other members stayed throughout the exercise. Among them, an equal number of men and women were present throughout the exercise.

The analysis shows: (i) presence of women and men nearly or in equal number; (ii) a sustained physical presence of participants throughout the exercise; and (iii) the influence of time over their participation. As the time for completing the exercise increased, the participants were likely to leave.

Table 5.5: Participation in Social Mapping at Chellakuttiur

No. of participants / *Duration (in minutes)*	*Participants at the beginning*			*Participants who joined latter*			*Participants who left*			*Participants at the end*		
	Male	*Female*	*Total*	*Male*	*Female*	*Total*	*Male*	*Female*	*Total*	*Male*	*Female*	*Total*
0 – 45	5	3	8	3	3	6	—	—	—	8	6	14
45 – 90	8	6	14	3	4	7	—	—	—	11	10	21
90 – 135	14	10	21	6	3	9	3	1	4	17	12	29
135 – 180	17	12	29	1	2	3	4	1	5	14	13	27
180 – 225	14	13	27	—	—	—	2	1	3	12	12	24
225 – 270	12	12	24	—	—	—	—	—	—	12	12	24
270 – 315	12	12	24	—	—	—	4	2	6	8	10	18
315 – 360	8	10	18	—	—	—	—	—	—	8	10	18

Total time taken = 360 minutes

Time spent on rapport building is not added here. Time spent is related to activities like drawing of the map, data collection, data analysis, reporting and sharing of data.

We have so far analyzed the physical presence of participants in the exercise. Now, we will examine the 'active participation of the participants in the exercise. In an exercise like mapping, every one cannot actively participate at a time. Some may participate in the form of 'gathering participants for mapping'; some may participate 'by gathering material for the exercise'; some may participate by actually 'drawing the map on the ground'; and some may participate by 'giving information' and so on. This would mean that all could not actively participate throughout the exercise. Yet, if a participant contributes in one form or the other in successful completion of the exercise, he/she may be termed as an active participant. Table 5.8 gives the extent of active participation at different stages of exercises. It may be noted that of the 69 participants from the three villages put together, around 74 per cent of the participants were active. Village-wise analysis shows that the percentage of active participants was found to be more in the case of smaller village. For instance, the percentage of active participants was 75 in Pannaikulam and 85 in Karuthakkapatti, whereas in Chellakuttiur, it was 65.

It could also be noted that women were found to be more active in general. For instance, of the total of 12 women participants in Chellakuttiur, 75 per cent were active; whereas the percentage of active men participants was 59. The same trend could be seen in other villages too.

Participation in different stages of PRA exercise indicates the involvement of women in all stages. However, they are found to have played a predominant role in gathering the participants and the material for mapping. In activities like 'mapping,' giving information and analysis of information, their participation was found to be nearly equivalent to men.

Conclusion

Participatory methods have been quickly spreading. They are applied in a wide variety of fields in different settings. The rapid spread and scaling up of these methods have raised certain questions and doubts on the efficacy of the methods. Studies have been undertaken to dispel such doubts on the efficacy of

Table 5.6: Participation in Social Mapping at Pannaikulam

No. of participants Duration (in minutes)	*Participants at the beginning*			*Participants who joined later*			*Participants who left*			*Participants at the end*		
	Male	*Female*	*Total*	*Male*	*Female*	*Total*	*Male*	*Female*	*Total*	*Male*	*Female*	*Total*
0 – 45	7	5	12	2	3	5	—	—	—	9	8	17
45 – 90	9	8	17	1	1	2	—	—	—	10	9	19
90 – 135	10	9	17	2	1	3	1	2	2	11	9	20
135 –180	11	9	20	—	—	—	2	1	3	9	8	17

Total time taken =165 minutes

Table 5.7: Participation in Social Mapping at Karuthakkapatti

No. of participants Duration (in minutes)	*Participants at the beginning*			*Participants who joined later*			*Participants who left*			*Participants at the end*		
	Male	*Female*	*Total*	*Male*	*Female*	*Total*	*Male*	*Female*	*Total*	*Male*	*Female*	*Total*
0 – 45	6	7	13	2	2	4	—	—	—	8	9	17
45 – 90	8	9	17	1	1	2	—	—	—	9	10	19
90 – 135	9	10	19	2	1	3	1	1	2	10	10	20

Total time taken 135 minutes

Table 5.8: Extent of Active Participation in Social Mapping

Details of data	*Chellakuttiur*			*Pannaikulam*			*Karuthakkapatti*			*Total*		
	Male	*Female*	*Total*	*Male*	*Female*	*Total*	*Male*	*Female*	*Total*	*Male*	*Female*	*Total*
Gathering the participants for mapping	3	4	7	2	1	3	1	4	5	6	9	15
Gathering the materials for mapping	1	3	4	2	1	3	2	4	6	5	8	13
Mapping	3	2	5	2	2	4	2	2	4	7	6	13
Giving information	4	3	7	3	2	5	3	3	6	10	8	18
Analysis of information	1	2	3	2	1	3	2	1	3	5	4	9
Reporting	1	1	2	2	—	2	1	1	2	4	2	6
Number of active participants	10	9	19	9	6	15	8	9	17	—	—	—
Passive participants	7	3	10	2	3	5	2	1	3	—	—	—

participatory methods. But such studies are sporadic in nature. Many practitioners and researchers emphasized the need to organise more such studies to test the efficacy of participatory methods. Hence, an attempt has been made to analyze the efficacy of participatory method at two levels (i) efficacy of participatory methods as envisaged by NGOs and (ii) assessing the efficacy of a method through an empirical field studies. The NGOs who have been practicing participatory methods in the field are of the view that participatory methods are quite effective in terms of 'reliability', 'accuracy', 'cost-effectiveness', 'participation', 'appropriateness', 'adequacy', 'simple to analyse', 'simple to practice'. However, they have reservations on efficacy in terms of 'rapidity'.

The empirical analysis of comparison of the data collected under 'participatory mapping' and 'survey method' has conclusively proved that a PRA method like participatory mapping help gathering, analyzing and reporting the data in a shorter period of time than the survey method. The time taken to complete a PRA, if we consider the 'community time', is most likely to exceed the time taken under survey methods. But then the community time need not be considered as the multiple benefits that the community members derive because it is likely to outweigh the cost in terms of the time they spent. PRA method is found to be 'cost-effective' as the cost of conducting survey to collect census data is more than double the cost of organizing PRA method. Regarding accuracy, there is very negligible variation in the data collected under the two methods. In some cases, the data collected through PRA are found to be more accurate whereas in some other cases the survey data is found to be accurate. The overall analysis shows that PRA data can be as accurate as survey data.

Participation in PRA is quite high and active. Contrary to some of the findings of the earlier studies, the women participated very actively. Their participation, if properly facilitated, would be better than the participation of men. The participatory methods, therefore, as perceived by the NGO and as inferred from the field study are proved to be effective. However, it must be noted that we have experimented with participatory mapping only for collection of quantitative data. But then, PRA's strength lies in the collection of qualitative data. PRA's efficacy in gathering, analysing and reporting of qualitative data is therefore, a potential and promising area for systematic research by social scientists.

NOTES AND REFERENCES

1. Robert Chambers, *Rural Appraisal: Rapid, Relaxed and Participatory,* Discussion Paper 311, IDS, Brighton, October 1992, p. 1.
2. Robert Chambers, "Participatory Rural Appraisal: Analysis of Experience", *World Development,* Vol. 22, No. 9, 1994, p. 1253.
3. Robert Chambers, "Making the Best of Going to Scale", *PLA Notes,* No. 24, October 1995, pp. 57-58.
4. Neela Mukherjee, *Participatory Rural Appraisal: Methodology* and *Applications,* Concept Publishing Company, New Delhi, 1994, p. 22.
5. David C.Korten, Social Science in the Service of Social Transformation in Cysthia C.Veneration (Ed.) *A Decade of Process Documentation Research: Reflection and Synthesis,* Institute of Philippine Culture, Ateneo de Manila University, Manila, 1989, p. 10.
6. *Ibid.,* p. 10.
7. See Araon Wildavsky, *"Speaking Truth to Power: The Art and Craft of Policy Analysis*", Little Brown, 1979. Also see John Friedman, *Planning in the Public Domain: From Knowledge to Action*, Princeton, New Jersey, Princeton University Press, 1987.
8. David C.Korten, Social Science in the Service of Social Transformation, *op.cit.*, p. 6.
9. Robert E.Rhoades, "Basic Field Techniques for Rapid Rural Appraisal", in *Proceedings of the 1985 International Conference on Rapid Rural Appraisal,* 1987, p. 114.
10. Robert Chambers, *Whose Reality Counts: Putting the First Last*, Intermediate Technology Publications, London, 1998.
11. Irene Guijt, Andrea Cornwall, "Critical Reflections on the Practice of PRA", *PLA Notes, No.24,* October 1995, p.2.
12. Robert Chambers, "Participatory Rural Appraisal: Challenges, Potentials and Paradigm", *World Development,* Vol. 22, No. 10, 1994, p. 1273.
13. *Ibid.*, p. 1273.
14. *Ibid.*, p. 1273.
15. *Ibid.*, p. 1273.
16. *Ibid.*, p. 1274.
17. David Mosse, *Authority, Gender and Knowledge: Theoretical Reflection on the Practice of Participatory Rural Appraisal*, Network Paper 44, The Overseas Development Administration, London, December 1994, p. 9.
18. The author (first) attended a three-day PRA training workshop organized by SPEECH, a voluntary organization in a village called Kadambankulam.

The trainers have been repeatedly emphasizing the concept of self-development and the need to reduce dependency culture among the participating villagers. On the final day during the valediction, the village leader stood up and said "we have been wholeheartedly participating in all the PRA exercises for the last three days as desired by you. What are you going to offer us in future?" This shows that the dependency syndrome is deep-seated in the minds of the villages. We cannot avoid this.

19. Bjorkman J.W, "Who governs the Health Sector", *Contemporary Politics,* 399, July, 1985.
20. Mehta *et al.,* "Gender issues in Participatory Natural Resource Development: A review of KRIBP Experiences", *KRIBP Working paper,* No. 7, Centre for Development Studies, University of Wales, Swansea.
21. Andrea Cornwall and Rachal Jewkes, "What is Participatory Research", *Social Sciences Medicine,* Vol. 41, No. 12, 1995, p. 1667.
22. Jules N. Pretty, *Regenerating Agriculture: Policies and Practices for Sustainability and Self-reliance,* Vikas Publishing House Pvt. Ltd. New Delhi, 1995 p.178.
23. Andrea Cornwall and Rachal Jewkes, *op. cit.*, p. 1671.
24. Robert Leurs "Current challenges facing Participatory Rural Appraisal" *Public Administration and Development*, Vo. 16, 1996. P. 70.
25. Robert Chambers, "Participatory Rural Appraisal: Analysis of Experience", *op. cit.*, p. 1265.
26. John Moris and James Copstake, *Qualitative Enquiry for Rural Development: A Review* International Technology Publications on behalf of ODI, London, 1983, p. 34.
27. Margaret, Buchanan, Smith, "Innovations Towards Rapid and Participatory Rural Appraisal", in *PRA and Agriculture Abstracts* by Steven Moris, June 1995, IDS, London. Also see David Mosse, Authority, Gender and Knowledge: Theoritical Reflections on Participatory Rural Appraisal, *Economic and Political Weekly*, Vol. XXX, No. 11, March 18, 1995, pp. 569-578.
28. Narayanasamy N., and Boraian M.P., "Community Participation in Big and Small Villages", *RRA Notes,* No. 18, June 1993, pp. 44-46.
29. Robert Chambers, *Whose Reality Counts? Putting the First Last,* Intermediate Technology Publications 1998, London, p. 122.
30. Neela Mukherjee, *Participatory Rural Appraisal and Questionnaire Survey*, Concept Publishing Company, New Delhi, 1997, p. 108.
31. *Ibid.*, p. 108
32. Hubert observes that as per his results, the formal survey required 5 to 7 times more working days as the informal survey and the time needed to publish the results was 3 times longer for the formal survey. See Gubert, Jean Pierre, "Comparing two methods for implementing farming systems surveys in the Imbo Region of Burundi, 11th FSR/E Symposium, October 1991, Michigan State University, Michigan. Rajarathinam et.al., have estimated that survey of 412 households have taken 776 person hours as

against 76 person hours in PRA. See Rajarathinam et.al., *Validating Wealth Ranking of PRA and Formal Survey in identifying the Rural Poor,* Rusha, Rusha Campus, Vellore, April 1992.

33. Micheal Richards, Jonathan Davies and William Cavendish, "Can PRA methods be used to collect economic data? A non-timber forest producer case study from Zimbabwe", *PLA Notes,* 36, October 1999, p. 38-39.
34. The PRA team of GRI facilitated modelling exercise in a Pillaiarnatham village. The water distribution system was depicted in the model. It was found that the Harijan colony with 40 households did not have even a single water tap, whereas each non-harijan street with 25 to 30 houses has two or three taps. This immediately created an awareness among the harijans and started demanding water taps in their colony. This immediately led to an action plan which we subsequently implemented. See Narayanasamy, N., "PRA Exercise in an Indian Village: A Retrospective Evaluation of the First Exposure to the Process", *RRA Notes*, 19, February, 1994, pp. 19-23.
35. Pretty Jules, N, *Participatory Inquiry and Agricultural Research,* IIED, London, 1993.
36. Gill, Gerry, O.K., "The Data's Lousy, But Its all we've Got (Being a Critique of Conventional Methods)", *Gate Keeper Series,* No.38, Sustainable Agricultural Programme, IIED, London, 1993, p. 13.
37. Chamber, Robert, Whose Reality Counts?, *op.cit.*, p.141.
38. *Ibid.*, pp. 141-144.
39. Britha Mikkelsen, *Methods for Development Work and Research*, Sage Publications, New Delhi, 1995, p. 62.
40. Locke, Edwin, "Toward a Theory of Task Motivation and Incentives", *Organisational Situation and Human Performance,* 3, 1968, pp. 157-89.
41. Frank Heller, Eugen Pusic, George Strauss and Bernhard Wilpert, *Organisational Participation: Myth and Reality,* Oxford University Press, 1998, p. 10.
42. Britha Mikkelsen, *op. cit.,* p. 63.

6

SUMMARY OF FINDINGS AND POLICY IMPLICATIONS

Introduction

Participatory Rural Appraisal has been spreading quite fast. Practiced by NGOs as a tool for appraisal in the beginning, PRA, is being extensively used at different stages of the project cycle. It is gaining currency and momentum. Universities have recognized PRA as a method of research; a tool for carrying out their outreach programmes. Government have resorted to these methods for enlisting and strengthening the participation of people in development projects. Such a quick spread of PRA also bristles with certain problems. PRA has been abused and has resulted in gradual degeneration. The critics of PRA quite often question the quality of output and outcome of the PRA.

PRA has witnessed more than a decade of experience and experiments. NGOs have extensively used PRA. It is quite an appropriate time that we take stock of the state-of-art of PRA among NGOs in Tamil Nadu who are mostly instrumental in spreading the concept of PRA. The current study is therefore, undertaken with the primary objectives of studying the state of art of PRA among NGOs in Tamil Nadu; problems and challenges in practicing participatory methods; and the efficacy of participatory methods with reference to select variables.

The study is empirical in nature, based on primary data. Survey, case study and participatory methods were used to gather information and data.

The universe of the study encompasses the NGOs practicing PRA in Tamil Nadu. There were 72 NGOs under the fold of Tamil Nadu Resource Team (PRA). A preliminary questionnaire was sent to all the 72 NGOs. Totally 54 NGOs responded, and all of them come under the purview of the study. For an in-depth case study, four NGOs viz., OUTREACH, Bangalore, SPEECH, Madurai, ASSEFA, Lathur, (T.N) and Women's Collective, Chennai were selected. The criteria followed in selecting the NGOs are age, experience in PRA, exposure to PRA in terms of training undergone and practice of participatory methods in development projects. The selection was made based on preliminary survey and subsequent interviews with all the 54 NGOs.

The tools employed to gather information - both qualitative and quantitative - are a short mailed questionnaire to gather basic information about NGOs practicing PRA, Focussed Group Discussion to know the state of art in PRA, Semi-Structured Interview with the people and field workers of NGOs, participation in and observation of PRA exercises, training programmes and meeting of the groups. To assess the efficacy of participatory methods, a comparison was attempted between survey and participatory techniques. Tools used for this purpose are Interview Schedule under conventional method and Participatory Mapping under participatory method. The study is largely qualitative in nature. Hence, descriptive statistics were used to analyse the data.

State of Art of PRA

PRA practitioners in Tamil Nadu have started a network known as TNRT, which comprises 72 NGOs. All these NGOs employ PRA techniques in their activities. Totally 54 NGOs responded to our survey. The inferences presented under the state-of-art-of-PRA are related to these 54 NGOs.

A vast majority of the NGOs surveyed were established either in 80s or 90s. The birth and evolution of a few NGOs, more or less synchronize with the genesis and growth of PRA. The main focus of these NGOs in terms of objectives are: awareness creation, empowerment of marginalized and women, promotion of credit and savings, and community organization. The major sectors where the NGOs work include women and development, income generation, livestock, health, sanitation, agriculture, forestry, education, child development, housing, protection of human rights, etc. The activities undertaken are well in tune with the objectives of the NGOs.

Many of the NGOs are bigger in size as 40 per cent of the NGOs have a jurisdiction of more than 50 villages. Some of them cover villages spread across more than two districts.

The NGOs under report, mostly work among marginalized section of the community like women, agricultural labourers, scheduled castes, and child labourers as stated in their objectives.

NGOs, by and large, work with the people through development/welfare projects. The project has three distinct stages viz., formulation, implementation and evaluation. Methods followed at different stages of the project decide and determine the degree of participation of people and also realization of the objectives. In the formulation of projects, all NGOs adopt Participatory Rural Appraisal. A few among them used survey and also Logical Framework Analysis along with PRA. However, PRA is found to be a predominant method used for project formulation. The main method used for implementing and evaluating the project is PRA.

The main problems encountered in formulating the projects, as reported by NGOs, are: paucity of secondary data, lack of skill in drafting the project proposals, and problem in organising and completing survey. The major problems faced in implementing the project as reported by NGOs are: want of support from the government and local bodies, factions and feuds in the village; and high degree of expectation from the people. The prime problems witnessed in monitoring their projects are: inability of the field workers, lack of adequate feed-back, lack of technical knowledge and guidance and lack of seriousness among the beneficiaries.

The NGOs have tried different methods in solving the problems they faced in project formulation, evaluation and monitoring. But the major method was found to be PRA. Around 61 per cent of the NGOs felt that repeated use of PRA helped them to solve many of the problems; around 44 per cent were of the view that use of PRA right from proposal stage help them to overcome many of the problems. The other measures are: combination of different methods, frequent visits and meetings, involving stakeholders at different stages of the project, and so on.

PRA is around ten years old in Tamil Nadu. The practice of PRA by NGOs indicates that they are in different age brackets in terms of experience. Around 22 per cent of NGOs have more than 8 years of experience, 46 per cent have 6 to 8 years of experience and the rest have 4 -6 years of experience.

Around 87 per cent of the NGOs have had more than one training. Around one-fourth has undergone training for more than five times. Many of them have undergone training sufficiently for a longer period in reputed NGOs which have good trainers/ facilitators in participatory methods.

Regarding the number of PRA methods used, around three-fourths of the NGOs use 5 to 10 methods in their work while 11 per cent resort to extensive range of methods. The most popular method among the NGOs is participatory mapping. The other methods with high scores are group discussion, seasonal calendar, ranking exercises and Venn diagram. The methods used by the NGOs indicate that they use a combination of three or four methods and do go in for sequencing. However, all the NGOs do not practice the concept of sequencing. Sufficiently trained and consistent practitioners follow sequencing of participatory methods. Further, sequencing depends on the type, nature and quality of data required by the participants.

PRA has several applications. They are being applied by the NGOs in many fields of activities. Prominent among them are women development (96.29%), micro-credit (87.03%), rural water supply and sanitation (83.33%), animal husbandry (83.33%), micro enterprises (74.07%) and health (74.07%). The other areas of application include agriculture, forestry, education and training.

PRA has been applied in areas like organization building and policy initiative by selected NGOs.

The benefits of PRA are varied. An interesting feature of the PRA is that all stakeholders derive benefits through the use of PRA. The benefits in general as reported by the NGOs are: perfect assessment of the situation at micro level; collection of wide range of information; building bridge between insiders and outsiders; cost effective; quicker; accuracy of data; simple to practice even by the layman; collection of qualitative data; creation of mutual respect and so on. An important criticism against PRA is that the data collected through PRA are not accurate. But observation of majority of the NGOs clearly indicate that PRA data can be accurate.

The benefits accruing to the people through PRA as reported by the NGOs are: self awareness and self assessment by people, free and frank articulation of needs, problems and aspirations, easy to understand and simple to practice, assessment of existing situation resulting in visualization of future plan, participation of passive members, active participation of the people in implementing the project, skill development in analysis and planning and participatory decision making.

NGOs practicing PRA do face problems. Problems are classified into four categories viz., (i) people-based; (ii) team-based; (iii) approach-based; and iv) tool-based.

The major people-based problems are: (i) high expectations from the people especially in the case of 'organised PRA'; (ii) PRAs are not equally accessible to all sections of the community especially in the heterogeneous community; (iii) domination of vested interests affect the quality and validity of the information; and (iv) less participation of women due to social constraints.

The 'team-based problems' are: (i) failure to constitute a good representative group of people from the village; (ii) leaving the poor; (iii) meeting the people through intermediaries (contact persons); (iv) inability to review and evaluate the process and outcome of PRA exercises; (v) tendency to practice the familiar PRA exercises and thus restricting the innovations in PRA (vi)absence or less number of women in PRA team; and (vii) failure to establish adequate rapport with the people.

The 'approach-based' problems as reported by the respondents are: (i) failure to conduct plenary session in the village resulting in sacrificing the principle of 'sharing'; (ii) neglect of passive participants; (iii) following the same approach in all the villages; (iv) wrong choice of venue; (v) failure to have 'night stay' in the village; (vi) lack of transparency on the part of outsiders; and (vii) lack of flexibility.

The 'tool-based' problems are: (i) difficulties in practicing visual exercises like mapping in a big village; (ii) difficulties in bringing out visual presentation in exercises like trend analysis, time line, livelihood analysis etc.; (iii) occupational diversity and heterogeneity make exercises like wealth ranking and seasonal calendar rather difficult and less meaningful; and (iv) problem in gathering and using locally available materials.

The major solution to many of the problems cited lies in continuous and consistent use of participatory methods and strict adherence to the process and philosophy both on and off the field.

Use of Participatory Methods

The NGOs emphasize collegiate type of participation in their development work-the core elements of such participation being process-orientation, bottom-up approach, and participatory methods. Participatory Rural Appraisal has offered abundant scope for NGOs to seek collegiate type of participation. NGOs do use participatory methods at different stages of the project cycle.

The experience of selected NGOs practicing Participatory Rural Appraisal methods have proved that they resort to PRA methods as these methods initiate, enhance and reinforce participation of the people. They consider PRA as less costly, less time-consuming and user-friendly in gathering information, in formulating projects and in executing and evaluating projects. Hence, they apply PRA at different stages of the project.

The NGOs view PRA as an effective methodology of gathering information, help listening to the voice of the hitherto neglected and unheard voices. They also view PRA as an effective methodology for formulating projects as they help in integrating the aspirations, needs and priorities of the people in their projects.

They consider PRA as an effective methodology for executing and evaluating their programmes as these methods facilitate active participation of people in implementing the activities and help in the assessment of their situation through periodical review and dialogue.

The case studies proved that PRA is quite relevant at different stages of the project cycle. PRA helps them to correctly assess the ground realities. Further, application of PRA at different stages of the project was not viewed in isolation. It is a continuum. Appraisal leads to formulation of projects and planning of activities; it further leads to implementation, monitoring and evaluation. The uses of methods facilitate the participants to think, analyse and act. NGOs, therefore, report that they never consider the use of PRA as irrelevant. On the other hand, each and every method they have used, has relevance to the field situation. The methods are also found to be effective at every stage of the project.

The people, who are the partners in development, are very much appreciative of the participatory methods. People feel that participatory methods enable the illiterates and marginalized to articulate and to act. The methods have helped them in building up capacity, especially among the weaker sections of the community at different stages of the project cycle. They also helped in upgrading their analytical skills. Continuous interaction through participatory methods like group discussion, guided dialogue, visual presentation and shared presentation have helped in enhancing their skills in reading and writing. People feel that participatory methods could develop effective partnership among the different stakeholders of development intervention. Participatory methods enabled the women to discuss the issues/problems that affect not only their families but also the village. The experience of NGOs have proved that participatory methods have released the poor from the shackles of dependency and led them to independency through starting of micro-enterprises. The people also felt that these methods are very simple to understand and easy to practice.

Participatory methods do empower the poor. The poor normally remain powerless and voiceless. Participatory methods provide a medium for the poor to articulate their problems, concerns and priorities. They help them to work as a group. The group work gives them strength. This strength provides a base for social and economic empowerment. Participatory methods have also helped in making the illiterate people as resource persons in the subjects like PRA, community organisation and self-help groups. These methods have created a great deal of confidence-among them, the confidence of 'we-can-also-do'.

NGOs have also succeeded in institutionalising the concept and philosophy of PRA. NGOs started practicing the basics of PRA like openness, transparency, non-domination, elimination of hierarchy and sharing in their day-to-day work.

NGOs have also attempted a few innovations in participatory methods and have succeeded. To mention a few: (i) ladder scale to asses the impact of a programme; (ii) 100 seeds scoring to evaluate a programme or to find out community's choice of tree species; (iii) anxiety map to know what would be the role of different members of the family, if an additional activity were introduced; (iv) census maps to show details such as different seeds, stones, and vegetables on matters such as *bindhis*. NGOs have also improvised the methods to suit local situation and condition. For instance, Venn diagrams which are normally used to ascertain relationship are also used to prioritise the problems and also to evaluate a programme. NGOs also effectively combine participatory methods with other methods for better results.

Efficacy of Participatory methods

The advocates and practitioners of PRA claim that participatory approach (i) empowers the marginalized communities; (ii) helps in pursuing community-based process of development (iii) seeks and enables the expression and integration of local diversities; and (iv) helps initiate participatory research. The critics of PRA are of the opinion that the results of PRA were biased, impressionistic and unreliable. The enquiry based on PRA is termed as indisciplined and sloppy. Rigour and accuracy are assumed. The poor are left out. All these criticisms require attention and indepth analysis. The data collected through PRA must fulfil the criteria of effective information supply like comprehensiveness, cost-effectiveness, timeliness, operational relevance, delivery of information to appropriate levels of management, and appropriateness to targets and goals.

The efficacy of participatory methods are, therefore, assessed with reference to the above criteria at the level of the field workers and practitioners of PRA and at the field itself.

Perception of NGOs: To test the efficacy of PRA at the level of field workers and practitioners, criteria have been evolved based on the review of literature and our own past experience. The criteria considered were: (i) reliability; (ii) accuracy; (iii) timeliness; (iv) appropriateness; (v) rapidity; (vi) participation; (vii) cost-effectiveness; (viii) simple to analyse; (ix) positive attitude and behaviour; (x) sharing; and (xi) easy to practice. The field workers and practitioners of PRA in five selected NGOs have been asked to assess the efficacy of participatory methods using scoring method (PRA method). The results of analysis indicate that the NGOs are of the view that participatory methods are quite effective in terms of reliability, accuracy, cost-effectiveness, participation, appropriateness, adequacy, simple to analyse and simple to practice.

Among the five NGOs, which were involved in assessing the efficacy of participatory methods, two NGOs viz., SPEECH and OUTREACH have assigned very high score of 96 and 90 per cent respectively. SEED and ASSEFA have assigned scores of 84 and 83 per cent respectively. RENAISSANCE has assigned the score of 73 per cent. There seems to be a relationship between exposure and training in participatory methods and efficacy of participatory methods. The NGOs like OUTREACH and SPEECH practice participatory methods for a longer period of time, practice participatory methods at every stage of the project cycle, exposed to a variety of methods and therefore perceived a high degree of effectiveness. SEED and ASSEFA practice participatory methods for quite a long period of time. But application of the concept of PRA is limited and restricted. Hence, their perception on the efficacy, though high, is not upto the level of OUTREACH and SPEECH. RENAISSANCE has applied the concept in appraisal alone. Hence, it is yet to get convinced of efficacy of participatory methods. The relationship between efficacy of participatory methods as perceived by NGOs and variables like experience, exposure, training etc., need further probe based on empirical analysis and evidence.

Efficacy of PRA as experimented: The efficacy of PRA at this level has been tested in comparison with other methods of data collection. The criteria chosen to analyse the efficacy are 'cost-effectiveness', 'rapidity', 'accuracy' and 'participation'. The first two criteria have been considered as PRA has evolved as cost-effective and rapid method. The last two criteria have been considered since critics; (i) largely question the accuracy of data based on PRA and (ii) raise doubts about the participation of different types of people in the PRA sittings.

The experiment was conducted in three different types of villages in terms of size, caste, structure of settlement and occupation. The purpose is to see that conducting experiments on the effectiveness of PRA vis-à-vis survey does not bring any biased result, as organizing PRA in a small, homogeneous village is likely to be very quick and brings high degree of participation. Hence, three different types of villages.

The methods chosen to compare the data are social mapping under PRA and survey under conventional method. The data collected under the two methods are: name of the head of the household, religion, caste, marital status, age of the family members, literacy status of the family members, occupation, type of houses, extent of land owned, possession of live stock. The same types of data were gathered under two methods.

Rapidity: The factors considered for the time involved in data collection, analysis and report are: time taken for conducting PRA/Survey, time taken for crosschecking data, time taken for data management and time taken for dissemination of data. The time involved to collect the data under two methods has clearly indicated that the survey has taken 299 per cent of time than PRA for all the three villages put together. It must be noted that the data collected under survey were not shared with the participants whereas data collected under PRA method were shared among the participants. If we consider the time taken for sharing the data under survey method, the total time taken under survey would be still more.

The average time taken to complete survey per household is 14.40 minutes whereas for the PRA it is 4.81 minutes for all the villages put together.

Village-wise analysis of the time taken indicates that survey consumed 300 per cent time than PRA. Yet there are variations in the time taken across the villages. The time taken for a bigger village (Chellakuttiur) per household is 15.50 minutes for survey and 5.04 minutes for PRA. The time taken for survey per household in small villages viz., Pannaikulam and Karuthakkapatti is 12.98 minutes and 12.53 minutes respectively. The time taken for PRA per household in these villages is 4.45 minutes and 4.50 minutes respectively. The variation in time taken is due to size and heterogeneous characters of the village. Bigger village brings problems like building rapport, enlisting participation of different kinds of people, finding the right type of key informants, which influence the duration of PRA. But these factors do not have much influence on conducting survey. Yet, we find variations in the time taken to conduct survey. This is mainly because of the prevalence of joint family system and big size of the family. For instance, the average size of the family in the big village is 7.5 whereas in other two villages it is 4.5 each.

Time taken per unit under different components like rapport building, data collection, data analysis, reporting and sharing indicates that data collection takes much time under survey and PRA followed by data analysis, rapport building, and reporting.

Time analysis must include both researchers' time and participants' time. Participants' time especially in PRA is not included for obvious reasons: (i) PRA is organized at their convenience; (ii) organized PRA in such a way that the participants have not lost their wages; (iii) PRA is mutual learning; (iv) PRA invariably brings solutions to some long-standing problems. Organizing PRA has several other benefits too.

Cost-effectiveness: Calculation of cost of data gathering under two methods was based on manpower expenses, printing and stationery and documentation. The total cost of conducting survey among 213 households was Rs.2,340/- whereas the cost of collecting the same data under PRA was only Rs.964/-. The cost of conducting survey for a set of census data under survey was 243 per cent more than the cost incurred on PRA for collecting the same data. Village-wise analysis of the cost data indicates that the average cost of conducting a survey and PRA was much higher in big villages than in small villages. The study has, therefore, proved that collection of census data through PRA is less costly than through survey.

Accuracy: Survey and PRA output have provided exactly the same results on the number of households and type of houses. Population data in a small village under these two methods exactly tally with each other. But in a big village, there is a variation in all the three data. Here, the survey data was found to be accurate. The data on landless was similar under two methods. But there are very negligible variations of data in the 'extent of land owned' between these two methods. The data on cattle population was more accurate under PRA method than survey method. The data on age and literacy have shown a slight variation under two methods. The survey data were more accurate. The overall conclusion is that PRA data (quantitative) can be as accurate as data collected through survey.

Participation: Participation includes both passive (physical presence) and active participation. The key informants have been observed in the PRA sittings for their active and passive participation. The observation has shown that (i) equal presence of women in all the three villages; (ii) sustained physical presence of participation throughout the exercise; (iii) influence of time on participation. In other words, when an exercise takes time for its completion, the interest of participants wanes and thus results in lesser number of participants especially at the fag end of the exercise. This points to the fact that participatory exercises need to be completed within 2 to 2 1/2 hours.

The active participation is observed with reference to gathering participants for mapping, gathering materials for the exercise, drawing the map on the ground, giving information, analysis of information, and reporting. Out of 69 participants from three villages, 74 per cent of them were active and participated in two or three activities mentioned above. Village-wise analysis indicates the percentage of active participants was more in smaller village than in bigger village. The percentage of active participants in Chellakuttiur (bigger village) was 65 whereas it was 85 in Karuthakkapatti and 75 in Pannaikulam.

Women were active in general. In Chellakuttiur, for instance, 75 per cent of women actively participated whereas the active participants among men was 59 per cent only. The women do participate at different stages of the PRA. Their participation was high when compared to men in gathering the participants and gathering of material. Their participation was equal with men in activities like actual mapping, giving information and analysis of information.

Efficacy of participatory method as perceived by NGOs and experimented in the field point to the fact that participatory methods are quite effective.

POLICY IMPLICATIONS

Participatory development aims at enabling individuals, groups and communities to plan, execute and manage development programmes that improve their living conditions. They often need technical, administrative, organizational, financial and legal support in their endeavour. Some of these supports lie in the public domain. They may be cumbersome or obstructive of development process, and therefore need change. The study conducted at the grassroots level using participatory methods among NGOs reveals that the effective use of participatory methods can help the people to manage development. We strongly feel that the rich experiences of NGOs need to be harvested and made use of by the government, panchayat raj institutions and universities and colleges in their development interventions for the economic and social betterment of the people. Based on intensive field studies among select NGOs in Tamil Nadu state, we propose the following recommendations:

NGOs submit different project proposals for funding to the different departments in both the state and central governments. Such proposals are normally based on data collected through survey/ conventional methods. They may now asked to present the proposals based on participatory appraisal and participatory planning. For instance, Council for People's Action and Rural Technology (CAPART) provides financial assistance to NGOs throughout the country for undertaking several activities in a variety of fields at the micro level. The funding agencies like CAPART may make it obligatory that NGOs submit all their project proposals based on participatory methods.

The government, of late, has started emphasizing the critical role of participation in the development process. The government has been implementing several programmes, welfare and development, in rural areas. Yet, the approach continues to the "top-down". The time has come to reverse the trend. The development officials in the government should be exposed to the concept of participatory approach. They must be made to spend major chunk of their time in the field with the people. Needs assessment, prioritization of needs, selection of participants (beneficiaries), implementation of activities and monitoring and evaluation should be on participatory basis. People must be facilitated to know the ills of dependency and the merits of self-development. The government must, therefore, integrate a participatory process in all its welfare and development schemes. This requires intensive training to all government officials, especially those officials working in field of rural development.

The concept of self-help is quickly catchingup in rural areas. These groups mostly concentrate on savings and credit. The credit, by and large, is deployed for meeting the consumption requirements of the group, as the members of the group experience problems in identifying production oriented, gainful employment opportunities. Participatory methods could be effectively used to ascertain commercially viable and technically feasible income generating programmes for such groups. This again requires training. Such training could be effective and productive if offered by NGOs who have experience both in training and fieldwork.

The NGOs, which prepare project proposals based on participatory approach, may be given preference in funding.

The panchayats, of late, are vested with several responsibilities. Their tasks go beyond promoting civic amenities and facilities in the rural areas. They need to bestow much attention on the integrated development of the village that include socio-economic dimensions too. The performance of these tasks depends on effective micro level planning involving all segments of population. PRA comes as an effective tool for this purpose. It can be used for analyzing the problems of people and for finding solutions and resources for solving the problems of people.

The Panchayat Raj functionaries, therefore, need to be imparted training in PRA based micro planning. The CAPART can lend a helping hand in providing financial support for organizing such training programmes. NGOs working within the jurisdiction of Panchayati Raj institutions can offer such type of training programmes.

When a government department is engaged in the formulation of a programme, the NGO, which has sufficient experience in PRA, may be associated. This will ensure incorporation of essential participatory features in the programme.

The educational institutions, especially the universities and colleges are yet to introduce participatory research in their curriculum. Participatory research with an in-built mechanism for participatory development would open-up several opportunities and challenges for the teachers and students in the field of research. It would also help them to practice the concept of "learning by doing".

Participatory methods can be introduced in programmes like National Service Scheme, Nehru Yuva Kendra etc. The participatory methods would help the volunteers and the officials to quickly assess the situation in the village and plan feasible programmes with the active support and participation of the people. PRA with lot of 'fun' elements would facilitate the students in formulating and executing meaningful programmes and activities during their stay in the villages.

IMPLICATIONS FOR FURTHER RESEARCH

The present study has proved that Participatory Rural Appraisal methods could play a very significant role in empowering the people at different stages of the project cycle. NGOs have extensively experimented with these methods and techniques at the micro-level and found them highly rewarding in the areas of information gathering, planning, implementing and evaluating development projects. Drawing lessons from NGOs experience, others including the government and donor agencies have started practicing the concept of PRA at micro as well as macro-level. What is the experience of the government? Do the officials of the government have seriously taken up the ideology of PRA? Have they grounded the concept in the field with all its essential principles and core philosophy? If 'so', what are their experiences? What do they view about these methods? Have the methods brought the desired results? If so, how? If not, 'why'? These issues may provide a base for doing research on PRA as practiced by Government.

PRA is mostly applied at the micro-level. How appropriate is PRA for macro-level planning or extensive micro-level planning? And what might be the consequences for flexibility and innovation? How can micro-level PRA processes feed into macro-level policy making? Can communities use PRA to negotiate with local government and other institutions? The mechanisms involved are relatively unexplored, yet vitally important. These issues invite an in-depth analysis.

There are studies of the long-term economic benefits of participation. But none has, as yet, looked at the costs and benefits of participation through the whole life cycle of the project. Cost benefit analysis of participation therefore, will be an interesting area of study.

An important strength of PRA is the 'use of visuals,' because with visuals, a whole set of questions and discussions arises which does not arise in the verbal. PRA menu has more than 40 methods and techniques. Can visuals be used in all these methods and techniques? If 'no' what are the problems in using visuals in such methods? What do people view about the use of visuals? These issues call for an in-depth study.

PRA helps in generating a great deal of qualitative data. The validity and trustworthiness of these data are questioned. A comparative study of the validity of qualitative data gathered through participatory methods and conventional methods like survey may help in testing the trustworthiness of PRA-based qualitative data.

Participation is a complex process. It has both qualitative and quantitative dimensions. Development of indicators to measure these two aspects would be an interesting area of research.

The present study has measured the efficacy of participatory methods by comparing the information collected through PRA method against a conventional method. The variables considered are limited and quantitative in nature. Such a study could be repeated by drawing more methods from PRA menu and by considering more number of variables-both qualitative and quantitative. Such a study would help evaluate whether PRA can supplement or complement or replace the conventional methods.

Conclusion

Non-Government organization with deep roots in the rural soil, working among the marginalized section of the community through a multitude of welfare and development projects deploy, in their day-to-day operations PRA as a method. Though brimming with prospects, and bristling with problems, PRA steadily anchored into the appraisal, planning, implementation, monitoring and evaluation processes of the NGOs, poised for promoting greater community participation.

With prior training, adequate practice and constant applications, the NGOs use a plethora of PRA methods in varied spheres resulting in a series of benefits as well as scores of problems. Yet, PRA is considered less costly, less time consuming and user-friendly in gathering information for project formulation, execution and evaluation. Most of all, PRA is perceived as an effective methodology as it helps in integrating the aspirations, needs and priorities of people in project formulation. PRA builds up people's capacity, upgrades their analytical skills, releases them from the shackles of dependency, facilitates greater independence through starting micro enterprises, mobilizes the marginalized people and empowers the poor and weak. Ultimately, NGOs have therefore institutionalized the concept of PRA – some have attempted innovations in PRA methods such as ladder score, 100 seeds scoring, anxiety map amd census map, while others have improvised the existing methods and integrated participatory methods with other methods for better results.

The advocates of PRA adore its benefits while its critics admonish saying that PRA is biased, impressionistic, unreliable, undisciplined and sloppy. An assessment of the efficacy of participatory methods using scoring method (PRA method) among the NGOs unfolds that participatory methods are quite effective in terms of criteria such as: reliability, accuracy, cost-effectiveness, participation, appropriateness, adequacy, simple to analyze and simple to practice. NGOs, with long experience, assigned a higher score of 90-96 per cent while those with experience gave a lower score of 73 to 83 per cent.

A comparative analysis of PRA and survey revealed that PRA took only about a third of survey time and cost, data were as accurate as under survey and the participants were active throughout the exercise under PRA, as against 'fag-end fatigue' in surveys.

The experience of NGOs practicing PRA and the comparative assessment of Survey and PRA proves beyond doubt that PRA is a unique method for universal application, which not only enjoys primacy but also commands supremacy over the survey method. PRA, is by all means, more efficacious, enabling and an empowering method while ensuring participatory development at the grassroots level.

BIBLIOGRAPHY

Allan Kaplan, (1999), *The Development of Capacity,* NGLS Development Dossier, UN Non-Governmental Liaison Service, Geneva.

Berg, Bruce L. (1989), *Qualitative Research Methods for the Social Sciences,* Allyn and Bacon.

Berlage, L. and O. Stokke (eds.) (1992), *Evaluating Development Assistance: Approaches and Methods,* EDI Book Series 14, Frank Cass.

Britha Mikkelsen, (1995), *Methods for Development Work and Research,* Sage Publications, New Delhi.

Campbell, J. Gabriel, Ramesh Shrestha and Linda Stone (1979), *The Use and Misuse of Social Science Research in Nepal,* Research Center for Nepal and Asian Studies, Tribhuvan University, Kirtipur, Kathmandu.

Casley, D.J. and Lury, D.A. (1981), *Data Collection in Developing Countries,* Clarendon.

Cernea, Michael M. (ed.) (1985), *Putting People First. Sociological Variables in Rural Development,* Johns Hopkins Press, Baltimore.

Chambers, Robert (1993) *Challenging the Professions: Frontiers for Rural Development,* Intermediate Technology Publications, London.

Chambers, Robert, Tushaar Shah and N.C. Saxena (1989), *To the Hands of the poor: Water and Trees,* Oxford and IBH, New Delhi.

Chambers, Robert (1987), *Rural Development - Putting the Last*

First, Longman Scientific and Technical, New York.

Chambers, Robert (1997), *Whose Reality Counts? Putting the First Last,* Intermediate Technology Publications, London.

Gill, Gerard, J. (1993), *OK, The Data's Lousy, But It's All We've Got (Being a Critique of Conventional Methods),* Gatekeeper Series 38, IIED, London.

Gardon R.Conway, Jennifer A. McCracken and Jules N. Pretty (1987), *Training Notes for Agro-eco-system Analysis and Rapid Rural Appraisal,* IIED, London.

Gow, D (1987), *Rapid Rural Appraisal: Social Science as Investigative Journalism,* in: Finsterusch, K., Ingersoll, J., and Liewellyn, L. (eds.) *Fitting Projects: Methods for Social Analysis for Projects in Developing Countries,* Lynne Rienner Publishers, Boulder, Colorado.

Jones, J., and Wallace, B., (eds.) (1985), *Social Sciences and Farming Systems Research: Methodological Perspectives on Agricultural Development,* Westview Press, Boulder, Colorado.

Khon Kaen University (1987), *Proceedings of the 1985 International Conference on Rapid Rural Appraisal,* Rural Systems Research and Farming Systems Research Projects, Khon Kaen, Thailand.

FAO (1989), *Community Forestry, Participatory Assessment, Monitoring and Evaluation,* Rome.

McCracken, Jennifer A., Jules N. Pretty and Gordon R. Conway (1988), *An Introduction to Rapid Rural Appraisal for Agricultural Development,* IIED, London.

Moser, C.A. and G. Kalton (1971), *Survey Methods in Social Investigation,* Heinemann Educational Books, London.

Mukherjee, Amitava (1995), *Participatory Rural Appraisal Methods and Application in Rural Planning,* Vikas Publishing House Pvt. Ltd., New Delhi. Revised edition published by Concept Publishing Company, 2004

Mukherjee, Neela (1994), *Participatory Rural Appraisal: Methodology and Applications,* Concept Publishing Company, New Delhi.

Mukherjee, Neela (1997), *Participatory Rural Appraisal and*

Questionnaire Survey: Comparative Field Experience and Methodological Innovations, Concept Publishing Company, New Delhi.

Narayanasamy, N., Dwaraki, B.R., Boraian, M.P., and Ramesh, R. (1998), *Analysing Community Problems: Tentacles of PRA Methodology,* Himalaya Publishing House, Mumbai.

Natarajan. I., et.al., (1993), *Comparative Study of Sample Survey and PRA Methodologies with special reference to Evaluation of National Programme on Improved Chulah,* NCAER, New Delhi.

Pretty, Jules N. (1995a), *Regenerating Agriculture: Policies and Practice for Sustainability and Self-reliance,* Earthscan Publications, London.

Pretty, Jules, Irene Guijt, John Thompson and Ian Scoones (1995), *A Trainer's Guide for Participatory Learning and Action,* IIED Participatory Methodology Series, IIED.

Rhoades, Robert. E. (1989), *Basic Field Techniques for Rapid Rural Appraisal,* KKU Proceedings, Thailand.

Rhoades, Robert E. (1979), *The Art of Informal Agricultural Survey,* Social Science Department Training Document, International Potato Centre, Lima, Peru.

Scoones, Ian and John Thompson (eds.) (1994), *Beyond Farmer First: Rural People's Knowledge, Agricultural Research and Extension Practice,* Intermediate Technology Publications, London.

Scrimshaw, Nevin and Gary R. Gleason (eds.) (1992), *Rapid Assessment Procedures: Qualitative Methodologies for Planning and Evaluation of Health Related Programmes,* International Nutrition Foundation for Developing Countries, Boston MA, USA.

Scott R.A., and Shore A.B. (1979), *Why Sociology does not Apply: A Study of Sociology in Public Planning,* Elsvien, New York.

Thompson, Paul (1998), *Voice of the Past: Oral History,* Oxford University Press, London.

Whyte, William Foote (1991), *Participatory Action Research,* Sage Publications, London.

ARTICLES

Absalom, Elkanah et al. (1995), "Participatory Methods and Approaches: Sharing our concerns and looking to the future", *PLA Notes,* 22, IIED, London, February.

Atkinson, S. "Interviews with Key Informants and Household Surveys: Central Ethiopia", *Disaster,* 16(3).

Belshaw, D. (1981), "A Theoretical Framework for Data-economizing Appraisal Procedures with Applications for Rural Development Planning", in Longhurst (ed.) *IDS Bulletin.*

Birgegard, Lars-Erik (1991), "A process Approach to Policy and Project Analysis", *PAPPA, SUAS,* Issue Paper No.13, Uppsala, February.

Buchanan-Smith, M. et al. (1993), "Finding Out How People Prioritise Their Food Security Problems in Chad: the challenges of RRA at national level", *RRA Notes,* 18, IIED, London, June.

Carruthers, Ian and Robert Chambers (1981), "Rapid Appraisal for Rural Development", *Agricultural Administration,* 8 (6).

Chambers, Robert (1981), "Rapid Rural Appraisal: rationale and repertoire", *Public Administration and Development,* 2 (2).

Chambers, Robert (1993), "Methods for Analysis by Farmers: The professional challenge", *Journal for Farming Systems Research-Extension,* 4 (1).

Chambers, Robert (1994), "The Origins and Practice of Participatory Rural Appraisal", *World Development,* 22 (7), July.

Chambers, Robert (1994), "Participatory Rural Appraisal (PRA): Analysis of Experience", *World Development,* 22 (9), September.

Chambers, Robert (1994), "Participatory Rural Appraisal (PRA): Challenges, Potentials and Paradigm", *World Development,* 22(10), October.

Conway G.R. (1988), "What is in a name?", Editorial, *RRA Notes,* 1, IIED, London, June.

Cornwall, Andrea and Sue Fleming (1995), "Context and Complexity: Anthropological reflections on PRA", *PLA Notes,* 24, IIED, London, October.

Cornwall, Andrea and Rachel Jewkes (1995), "Participatory Research", *Social Science and Medicine,* 41(12).

Davis, Robert (1997), "Combining Rapid Appraisal with Quantitative Methods: An Example from Mauritania", *PLA Notes,* 28, IIED, London, February.

Gill, Gerad J. (1991), "But How Does It Compare with REAL Data?", *PRA Notes,*14, IIED, London, December.

Gill, Gerard J. (1993), "Are Some 'Participatory' Techniques Culturally Biased? (Or: Are We Hooked on Mom's Apple Pie?)", *PRA Notes,* 18, IIED, London, June.

Guijt, Irene (1994), "Making a Difference: Integrating gender analysis into PRA training", *PRA Notes,* 19, IIED, London, February.

Guijt, Irene and Andrea Cornwall (1995), "Editorial: Critical Reflections on the Practice of PRA", *PLA Notes,* 24, IIED, London,October.

Hari M.John, (1992), "In True Partnership with people: Participatory Action Research and the Poor", *SEARCH News,* 7(1), January-March.

Inglis, Andy (1991), "Harvesting Local Forests Knowledge: A Comparison of PRA and Conventional Surveys", *RRA Notes,* 12, IIED, London, July.

Jodha, N.S. (1998), "Poverty Debate in India: a Minority View", *Economic and Political Weekly,* Special Number: 2421-28, November.

Katherin Schreckenberg, (1995), "The Respective Merits of PRA and Conventional Methods for Long-Term Research", *PLA Notes,* 24, IIED, London, October.

Lindberg.C., Laiske, V.M., Ostberg. N., and Mungiongo, C., (1995), "Handle with Care! Rapid Studies of the Poor", *PLA Notes,* 22, IIED, London, February.

Margaret Buchanan Smith, et.al., (1993), "Finding out how people prioritize their Food Security Problem in Chad: The Challenges of PRA at National level", *RRA Notes,* 18, IIED, London, June.

Mascarenhas, James, (1991), "PRA and Participatory Learning Methods: Recent Experiences from MYRADA in South India", *RRA Notes,* 13, IIED and MYRADA, London, August.

Mascarenhas, J., Parmesh Shah, Sam Joseph, Ravi Jayakaran, John Devaram, Vidya Ramachandran, Alosiyus Fernandez, Robert Chambers and Jules Pretty, (1991), "PRA in India: Review and Future Directions", *RRA Notes,* 13, IIED and MYRADA, London, August.

Mascarenhas, James (1991), "PRA and Participatory Learning Methods: Recent Experiences from MYRADA and South India", *PRA Notes,* 13, IIED, London, August.

Mascarenhas, J. and P.D. Prem Kumar (1991), "Participatory Mapping and Modelling: User's Notes", *RRA Notes,* 12, IIED, London, July.

Marian Meyar and Naresh Singh, (1997), "Two Approaches to Evaluating the Outcomes of Development Project", *Development in Practice,* 7 (1), February.

Michael Richards, Jonathan Davies and William Cavendish, (1999), "Can PRA Methods be Used to Collect Economic Data?: A Non-Timber Forest Product Case Study from Zimbabwe", *PLA Notes,* 36, IIED, London, October.

Michele Nori, Adane Hirpa and Giovanni A. Ferrai, (1999), "Complementary Methods to Understand Land Use Changes: An Example From the Ethiopian Rifts Valley", *PLA Notes,* 35, IIED, London, June.

Mosse, David (1995),"Social Analysis in Participatory Rural Development", *PLA Notes,* 24, IIED, London, October.

Mosse, David (1995), "Authority, Gender and Knowledge: Theoretical Reflection on Participatory Rural Appraisal", *Economic and Political Weekly,* 20(11), March 18.

Mukherjee, Neela, (1992), "Villagers' Perception of Rural Poverty

Through the Mapping Methods of PRA", *RRA Notes,* 15, IIED, London, May.

Richards, Paul (1995), "Participatory Rural Appraisal: A quick and dirty Critique", *PLA Notes,* 24, IIED, London, October.

Robert Leurs, (1996), "Current Challenges Facing Participatory Rural Appraisal", *Public Administration and Development,* 16.

Rosemary McGee, (1997), "Ethnography and Rapid Appraisal in Doctoral Research on Poverty", *PLA Notes,* 28, IIED, London, February.

Sarah Gammage, (1997), "PRA and its Complementarities with Household Survey Methodologies", *PLA Notes,* 28, IIED, London, February.

Scoones.I, (1995), "Investigating Difference: Application of Wealth Ranking and Household Survey Approaches among Farming Households in Southern Zimbabwe", *Development and Change,* 26.

Shah, Parmesh and Meera Kaul Shah (1995), "Participatory Methods Precipitating or avoiding conflict?", *PLA Notes,* 24, IIED, London, October.

Swift, J. (1981), "Rapid Appraisal and Cost-effective Research in West Africa", *Agricultural Administration,* 8, November.

Welbourn, Alice (1991), "The Analysis of Difference", *RRA Notes,* 14, IIED, London, December.

Welbourn, Alice (1992), "Rapid Rural Appraisal, Gender and Health-Alternative Ways of Listening to Needs", *IDS Bulletin,* 23(1).

William, Murphy (1992), "Creating the Appearance of Consensus in Mende Political Discourse" *American Anthropologist* (1).

REPORTS

Action Aid/IIED, (1992), "From Input to Impact: Participatory Rural Appraisal for Action Aid the Gambia", March.

Action Aid/IIED, (1989), "An Experiment with RRA in Ethiopia", October.

Assefa, Taye and Alemayehu Konde (eds.) (1996) "Proceedings of the National Conference on PRA", 12-15 February 1996, Addis Ababa, Farm-Africa, April.

Attwood, Heidi (1996), "South African Participatory Poverty Assessment Process: Were the Voices of the Poor Heard?" Paper for the PRA and Policy Workshop, IDS, Sussex, 13-14, May.

Beebe, J. (1985), "Rapid Rural Appraisal: The Critical First Step in a Farming Systems Approach to Research," Networking Paper No.5, Farming Systems Support Project, University of Florida, Florida.

Biggs, S. "Resource-Poor Farmers' Participation in Research: A Synthesis of Experience from Nine National Agricultural Research Systems", OFCUE Comparative Study Paper 3, International Service for National Agricultural Research, The Hague, 1989.

Bilgi, Meena, (1992), "Entering Women's World Through Men's Eyes—A Participatory Rural Appraisal at Boripitha," Netrang, Aga Khan Rural Support Programme, India, October.

Chambers, Robert, "Rural Appraisal: Rapid, Relaxed and Participatory", Institute for Development Studies, Discussion Paper No.311, Brighton, 1992.

Chambers, Robert (1974), "Managing Rural Development: Ideas and Experience from East Africa", Scandinavian Institute of African Studies, Uppsala.

Chambers, Robert (1995), "Poverty and Livelihoods: Whose Reality Counts?" IDS Discussion Paper 347, IDS, University of Sussex, January.

Guijt, Irene, and Jules N. Pretty (eds.) (1992), "Participatory Rural Appraisal for Farmer Participatory Research in Punjab, Pakistan", IIED, London, September.

IDS (1979), "Whose Knowledge Counts?" *IDS Bulletin,* 10(2).

IDS *Agriculture* (1996), Agriculture Pack, PRA, IDS, Sussex.

IDS Behaviour and Attitudes (1996), "Behaviour and Attitudes Pack", PRA, IDS, Sussex.

IDS Gender (1996) "Gender Pack", PRA, IDS, Sussex.

IDS People and Parks (1996), "People and Parks Pack", PRA, IDS, Sussex.

McCracken, J.A. (1988), "Participatory Rapid Rural Appraisal in Gujarat: a Trial Model for the Aga Khan Rural Support Programme" (India), IIED. London, November.

MYRADA, (1990), "PRA-PALM Series", Bangalore.

National Council of Applied Economic Research, (1992), "Evaluating the Impact of commercialization of Rural Dairy Sector: An application of RRA Techniques", New Delhi.

PRA Unit, Gandhigram Rural Institute, (1995), "PRA Retreat in a Rural Setting Gandhigram", Gandhigram, Tamil Nadu.

PRA Unit,(1996) "PRA at GRI: Looking Back at the Trodden Path", Gandhigram Rural Institute, Gandhigram.

PRA Unit, Gandhigram Rural Institute, (1997), "Hunger Free Area Programmes: A Pilot Study", Gandhigram, Tamil Nadu.

Robert E. Rhoads, "Basic Field Techniques for Rapid Rural Appraisal", Proceedings of the International Conference on Rapid Rural Appraisal, 1987.

Rajaratnam, Jolly, Ganeshan, C., Tharian, Helen, Babu and Abel. (1992), "Validating Wealth Ranking of PRA and Formal Survey in Identifying the Rural Poor", RUSHA Department, Christian Medical College & Hospital, RUSHA Campus, Vellore (TN), April.

Ramachandran, Vidya (1990), "A Workshop on Participatory Learning Methods, 8 to 12 January 1990", MYRADA Talavadi Project, PRA/PALM Series No.1, MYRADA, Bangalore.

World Bank (1992), "Effective Implementation: Key to Development Impact", Report of the World Bank Portfolio Management Task Force (Wapenhans Report), World Bank, Washington, October.

World Bank (1988), "Rural Development: World Bank Experience 1965-86", Operations Evaluation Department, The World Bank, Washington DC.

World Bank (1994), "The World Bank and Participation", Operations Policy Department, World Bank, September.

World Bank (1995), "World Bank Participation Source book", Environment Department Papers, World Bank, June.

INDEX

Accuracy, 151-54, 161-62, 174, 186-89, 195
Active participation, 189
Activity profiles, 15
Adequacy, 152
Adult literacy, 47
Aga Khan Rural Support Programme (Pakistan), 6
Agricultural Engineering Department, advantages of, 124-25
Agricultural labourers, 180
Agricultural research team, 7
Agricultural and animal husbandry, 128
Agro-eco-system analysis, 5-7, 8-9
Agro-service centres, 97-99
  problems in, 101-02
Animal husbandry, 128-29, 135, 181
Anxiety map through family counselling technique, 116-18, 185
Applied anthropology, 7
Approaches, 9-10
Approach based problems, 77-79, 183
Appropriateness, 151-52, 174, 186
Artificial shrimp of salt ponds, 27
ASA, Tiruchi, 65-85
Association for Sarva Seva Farms (ASSEFA). Lathur, 41, 85, 127-40, 144, 154, 179, 186
  Impact assessment through social mapping, 129-35
  participation empowers, 136-39
  project area integrated Rural Development Project ASSEFA, Sivaganga, 137-38
  Project formation through PRA, 128-29
Assessment, 182-84, 191-95
Assimilation, 38
Atkinson, 31
Awareness creation, 180

Benefit to people, 72
Biggs, S., 88
Bindhis, 125, 185
Bio-diversity, 27
Bio-intensive garden, 119
Bottom-up approach, 122
  advantages of, 89-90
Bottom-up micro planning, 90
Brainstorming technique, 110

Calves, 163
Card-Sorting method, 102
Caste composition, 155
Cattle population, 163
Cavendish, Willian, 24
Census map, 185
Challenges, 39, 148-49, 150-51
Chambers, Robert, 9, 67, 113, 148, 150, 160, 161
Chappathi diagram, 93, 104, 110, 113
  ranking of bad qualities in a family, 112
Checkdam, 97, 121
Chellukuthiur, participation in social mapping, 169
Child development, 180
Child labour, 119
Children's education, 47
Christian Aid, 141
Circle diagram, 17
Code of conduct, 140

'Collegiate' type of participation, 89
Common property resources, 125
Community organization, 180, 185
Comparison of data callected through social mapping and survey method, 164-65
Control-oriented administrative structure, 146
Conventional approach, 145-46
Conventional research methods, 30
Conventional sources, 163
Conway, Garden, 6
Cost of data gathering—comparison between survey and PRA method, 162
Cost-effectiveness, 25, 152-54, 160-61, 174, 185, 187-89, 195
Council for People's Action and Rural Technology (CAPART), 191, 192
Craft women PRA group, 25
Criticisms, 22, 147-48, 149-51, 182-85
Cropping pattern, 7
Cross-checking, 36
    data, 66, 157, 187-88

Dangers, 148-49, 150-52
Data collection, 10, 27-28, 35-40, 159-60, 166-68, 188-90
    methods, 155-56, 187-88
    tools for, 42-43
Davies, Jonathan, 24
Davis, Robert, 25
Deforestation, 27
Development challenges, 2-4
Development process, 2-3
    programmes, 146-47
Direct Matrix Ranking (DMR), 17
Domestic stakeholders, 149-50
Donor agencies, 62, 89, 193
Drinking water, 121
Duration of training, 64

Ecological degradation, 140
Efficacy, 40
    criteria, 151
Efficacy of participatory method, 174
Efficacy of PRA as experimented, 154-55, 187-88
Efficacy of PRA as reported by NGOs, 153
Eighth Five Year Plan, 88
Emerging and refugees, 47
Empowerment, 118
Empowerment of common people, 90
Enumerator training, 25

Fag-end fatigue, 195
Ferror, Giovanni A., 30
Field-based participatory training, 63
Field-oriented methods, 22
Flag of necessary, 22, 62
Flexibility, 11, 77-80, 150
Focused Group Discussion, 58
Food security, 31
    planning approach, 37
Forest degradation, 93
Forest management, 115
Formal questionnaire, 8
    surveys, 32
Framework of analysis, 43
Fuelwood trading system, 32
Funding problems, 140

Gammage, Sarah, 26
Gandhigram Rural Institute, 47, 151
Ganga Mahabir Mandram, 104
Gathering participation for mapping, 170
Gender differences, 25
Gill, G., 32
Giving information, 170
Gram Sabha, 127, 128, 135
    using participatory method, 128
Grey area, 109
Groundnut cropping pattern, 108
Group interviews, 16
Guided dialogue technique, 99

Handicapped, 126
Hirpa, Adane, 30

Hunger Free Area Programme, 28
Hypotheses, 40

Immunization status, 126
Impact assessment, 135
Impact assessment through social mapping, 129-35
Impact diagram, 113
ladder scale, 114
Implications for further research, 193-95
Income generation programmes, 50, 191
Inequality, 146
Inferences, 91-92, 118-19, 126-27, 139-40, 143-44
Informal Survey or Sondeo, 7-8
Informal survey to PRA, contribution of, 8-9
Information, cross-checking of, 150
gathering for planning, 89-90, 193-94
topics, 31
Inglis, Andy, 31
Innovations, 113-18, 125-26
Investigation, types, 31-32
Instrumental participation, 167
Interviewing map, 126
Irrigation facilities, 121

Joint family system, 155, 188-89
Joint Forest Management (JFM), 113, 114, 115

Kamakshi Mahabir Sangam, self-monitoring of, 109-10
Karuthakkapathi, participation is social mapping, 172
Keelneerpallam, Cheyyur taluk, 129
Key informant interviews, 16, 31-32
Knowledge for Action, 89

Ladder scale programme, 185
Landuse map, 13-14
Lathur Project, 127, 136
Lindberg, C., 32
Local management systems and resources, inventory, of 16-17
Local researchers and village analysis, 15
Local resource management, 119
Logging activities, 27
Logistic Framework Analysis (LFA), 58, 59, 129-34, 139-40, 180-81
Loiske, V.M., 32

Mani, N.D., 28
Manpower planning, 46
Matrices, 16-17
Matrix ranking, 17, 93
Measures taken to overcome the problems, 59-60
McGee, Rosemary, 29
Micro-credit, 181
Migration, 14
Mobility map, 97, 98
Mosse, D., 33
Mukherjee, Neela, 34
Mungongo, C., 32
MYRADA, Bangalore, 65, 85

Nandimangalam—advantage and limitation, 99-100
resource location and mobility map, 99
Narayanasamy, N., 28
Natarajan, I., 33
National Council of Applied Economic Research, 36
National Food Security System, development of, 37-38
National Service Scheme, 192
Nehru Yuva Kendra, 192
Night Halt, 18
Non-domination, 77
Non-Governmental Organizations (NGOs), 40-42
clients/beneficiaries of, 55-56
efficacy as perceived by, 152-54
fields of activities, 181

innovation in participatory methods, 185
jurisdiction of, 52-53, 55
method of operation, 56-57
objectives of, 49-52
in participatory methods, 39-40, 185-86
problems faced, 58-59
profile of, 48-50
response of, 48
sectorwise activities, undertaken by, 50-54
selection of, 41-42
their evolution, 48
tool based problems, 80-81
years of service, 48-49
NGOs practicing PRA, 183, 195
problems of, 182-83
Nori, Michael, 30

Objectives, 39-40, 49-52
Observation method, 42
Occuption, 155-56
ODA-funded natural resource project, 33
Off-setting biases, 12
Om Sakthi Mahabir Mandram (Women's Self-Help Group), 97, 101
On the spot analysis, 12-13
One off training, 63
Operation flood areas, 36-37
Operationalization of the concept, 40-41
Optimizing trade off, 12
Organisational analysis, 47
Ostberg, W.N., 32
OUTREACH, Bangalore, 41, 65, 85, 91-119, 144, 152-54, 179-80, 186
Inferences, 118-19
Innovations, 113-18
participatory empowers, 102-104
participatory methods, 118-19
fecilitated action, 97-99
participatory training, 110-18
programmes, 178-79
self-monitoring, 107-109
sequencing of methods and action, 104-107
training programmes, 100-101
using ladder scale, 113
using scoring method, 107
Usthanpalli, continuum between appraisal and planning, 93-97

Panchayat Budgets, 125
Panchayat Raj functionaries, 192
Panchayat responsibilities, 191-92
Pannaikulam, participation in social mapping, 171
Paraikulam experience, PRA in project cycle, 119-22
Participation, 166-68, 174, 189-90
alternative uses of, 167-68
empowers of, 102-104, 136-40
sex-wise, 168
Participation of people in projects, reasons of, 166
Participatory Action Research (PAR), 3-4
system of, 8-9
Participatory approaches, 3-10
agro-eco-system analysis, 5-7
applied anthroplogy, 7-8
informal survey or sources, 7-8
rapid rural appraisal, 8-10
Participatory community, 4-6
Patricipatory mapping, 163-64, 174
advantages of, 160-62
Participatory methods, 40, 191-92
advantages, of 89-90
applied areas of, 68
efficacy of, 145-174, 178-80, 185-86, 190, 195
powers of, 89-90
use of, 62-82, 183-84
Participatory methods benefit village panchayat, 125-27
Participatory methods by NGOs, use of, 67-68

Participatory methods in Project cycle, 88-144
Participatory monitoring and evolution, 47
Participatory Rural Appraisal (PRA), 10-12
  advent of, 10-12
  approaches, 10
  based qualitative data, trustworthiness, 194
  basic components of, 11
  banefits of using, 69-71
  conventional methods, 21-22
  description of, 145-46
  emphasizes qualities, 77
  evolution and growth of, 3-4
  experience in, 62-63
  factors of, 160-61
  high expectations from the people, 74
  methods of, 13, 24-25, 65-66, 151-52, 193-95
    used by NGOs, 66-68
  practitioners and literature on participation gains of, 90-92
  principles of, 11-12
  problems in practicing, categories of, 72-73
  state of art of, 46-80
  superiority of, 150-51
  techniques, 142-43
  training in, 63-64
  wealth ranking techniques of, 35
PRA in rehabilitation minor irrigation tanks, 122-25
Participatory training methods, 110-18
  programmes, 42-43
People based problems, 74, 75, 182
People old and young, categories, of, 167
Perception of NGOs, 186
Planting tree seedlings, 97
Plenary presentation, 110, 113
Policy implecations, 190-92
Poverty, phenomena of, 146-47
Power to the people, 146
Practical applications, 68-69
Practicing participatory methods, problems and challenges of, 178-80
Problems faced in monitoring and evaluation of the Project, 60-61
Problems and challenges, 178-80
Project cycle, 184
Projectized resource allocation schemes, 146
Promotion of credit and saving, 180

Qualitative information, 38
Qualitative and quantitative data, 27-28, 129-30
  dimensions, 194-95
Questionnaire survey, 34, 150

Rajaratnam, 35
Ramesh, R., 28
Rapid Rural Appraisal, 8-10, 148-50
Rapidity, 152, 154, 156, 186-87
Rahabilitating minor irrigation tanks, 122-25
Reliability, 152, 154, 174-75, 186-87, 195
RENAISSANCE, 154, 186
Repayment of IFAD Loan, 101
Report structure, 43
Resource map, 14
Reversal of learning, 11-12
Richards, Michael, 24, 159
Risk management strategies, 26
Rural development tourism, 150
  anti-poverty biases of, 9
Rural water supply and sanitation, 181

Sangam, 99-100, 101-102
Sanitation, 129
Sarva Seva Dairy Farm (SSDF), 136
Savings and credit, 119, 128, 129
Schreckenberg, Katherin, 29
Scoones, I., 34
Scoring method, 107

SEARCH, Bangalore, 65, 85
Seasonal calendar, 14, 65, 94, 104, 123, 181
SEED, 154, 186
Seeds scoring, 93, 115-16, 144, 185
  in community choice of species, 115
Seed Scoring method, ranking of bad qualities in a family, 110-11
Seeking diversity, 12
Self-correcting notes and diaries, 18
Self-critical awareness, 12
Self-evaluation, 102
Self-Helf Group (SHG), 97, 99-100, 102-104, 141, 185
  non-repayment of loan to, 104
Self-monitoring, 107-109
Semi-structural interviewing, 18, 179-80
Sequencing, 68
  of methods and action, 104-107
  of participatory methods, 66
SEVI, Kulithalai, 47
Sexual and reproductive health, 47
Shared presentation, 18
Sharing, 13, 152
  information, 14, 18
Shifting paradigm of development, 22
Siltation areas, 27
Smith, Buchanan, 37
Social constraints, 75
Social mapping, 14, 68, 121, 129, 155
  extent of active participation in, 173-74
Socio-economic development programme, Lathur Block, 130-34
Soil erosion, 93
Solution to problems, 81-84
  approach based problems, 81-82
  people based problems, 82-83
  team based problems, 83-84
  tool based problems, 84
Source of training, 64-65
SPEECH, Madurai, 41, 47, 65, 85, 152, 154, 179, 186
  adopted rapport building methods, 121
  advantages, 124-25
  development strategy of, 120
  inferences, 126-27
  mapping method, 144
  participatory methods benefit village panchayat, 125-27
    used by, 119-20
  PRA in Project cycle, Paraikulam experience, 119-22
  PRA in rehabilitation minor irrigation tanks, 122-25
  training programmes, 125
  using participatory methods, 119-20, 125-26
Spread of PRA among NGOs, 147-48
Stakeholder's participation, 23
  problems of, 89-90
Standardization of Minor Irrigation Tanks (SMIT), 123
State of Art of PRA, 62, 178-80
Sub-divided participation, types of, 88-89
Sub-Saharan Africa, agro-pastoralism in, 30
Sudagondapalli Banga Mahabir Sangam, expenditure in a year, 105-106
Sustainable manpower management strategy, 27
SWOT analysis, 65, 99

Tamil Nadu Agricultural University (Coinbatore), 47
Tamil Nadu Resource Team (TNRT), 41, 47, 65, 85, 179
Tamil Nadu, NGOs practicing PRA in, 178-79
Tamil Nadu Women's collective, Chennai, 47, 48, 65, 85, 140-44
  inferences of, 143-44
  objectives of, 141-42
Tank-bed land, 123

Tank rehabilitation/modernization, 119
Team based problems, 76-77, 182
Thailand, 6
Time line, 14, 121
Timeliness, 151, 186
Time taken analysis, 188
Time taken together data, a comparison between survey and PRA methods, 158
Tool based problems, 78-79, 183-84
Top-down approach, 122, 191
    assumption and practices, 1-2
Trade unions, 46
Training programmes, 64, 110-13, 179-80
    types of, 192
Transformation participation, 167
Transparency, 77, 143
Transect, 27, 121
Transect walk, 123
Transect zones, 14
Tree species, 185
Trend analysis, 14-15
Trend change, 93-94, 121-23, 139-40
Trend line/diagram, 15
Triangulation, 12
Trustworthiness, 161

Unemployment, 146
University of Chiang Mai, 6
Use of visuals, 193
Using participatory resource map, 122
Using participatory mapping, 135
Usthanapalli, continuum between appraisal and planning, 93-97
    seasonal calendar, 96
    trends, 95
    Venn diagram, 94

Vadapalai village, SMIT programme in, 123
Values addition, 107
Variables, 193-94
Venn diagram, 17, 65, 93, 119, 129, 144, 181, 185
Village Forest Committee, participation of, 115
Village organisation-linkage, 127
Visual literacy, 79
Visual teachniques, 27
Voluntary Organizations Women's Staff Association, 140

Water table fluctuation, 27
Watershed management, 119
Wealth ranking, 17, 65
Welfare and development schemes, 191
Women's association, 136
Women's collective, Chennai, 85, 179
Women development, 128-29, 181-82
Women perspectives, articulation of, 149-50
Women Sangam, 141
World Bank, 46

Zimbabwe, 34